Inventive Life:
Approaches to the New Vitalism

Edited by
Mariam Fraser, Sarah Kember and Celia Lury

Inventive Life:
Approaches to the New Vitalism

Edited by
Mariam Fraser, Sarah Kember and Celia Lury

Contents

Originally published as Volume 22, Number 1 of *Theory, Culture & Society*

First published 2006

© Mariam Fraser, Sarah Kember and Celia Lury 2006

Published in association with *Theory, Culture & Society*, Nottingham Trent University

SAGE Publications Ltd
1 Oliver's Yard
55 City Road
London EC1Y 1SP

SAGE Publications Inc.
2455 Teller Road
Thousand Oaks, California 91320

SAGE Publications India Pvt Ltd
B-42, Panchsheel Enclave
Post Box 4109
New Delhi 110 017

British Library Cataloguing in Publication data

A catalogue record for this book is available from the British Library

ISBN 1 4129 2036 1

Library of Congress Control Number Available

Typeset by Type Study, Scarborough, North Yorkshire
Printed on paper from sustainable resources
Printed in Great Britain by Athenaeum Press, Gateshead

Inventive Life
Approaches to the New Vitalism

Mariam Fraser, Sarah Kember and
Celia Lury

[I]n vain we force the living into this or that one of our moulds. All the moulds crack. They are too narrow, above all too rigid, for what we try to put into them. (Bergson, 1911: x)

THIS INTRODUCTION addresses how and why vitalism – the idea, originating in the 18th and 19th centuries, that life cannot be explained by the principles of mechanism – matters now. It does so by considering what we will call vital processes. One aim is thus to think about process, that is, what is distinctive about process as a mode of being. A second aim is to address some of the ways in which attempts are currently being made to introduce information, knowledge or 'mind' into social and natural entities, making them less inert, more process-like: bringing them alive. The two aims are held in tension. Thus while we consider the specific, contemporary set of circumstances in which the vitality of (social and natural) processes is currently being proposed – namely, the introduction of understandings of information, complexity, and cybernetics in the economy, science, and art – this is set alongside historical and philosophical understandings of process. The effect of this double focus, we hope, is to enable us to pose questions about the role and status of (thinking) life across the natural sciences, the social sciences and the humanities. These questions return us to a consideration of what might be included in the life sciences, how such sciences might be conducted, and who conducts them. This consideration does not take the form of a critique from without or a commentary from above, but emerges from within. Our hope is that a concern with vital processes will enable us to think about change – both novelty and endurance – in a world which 'might be different but is not' (Haraway, 1997: 97).

In the article 'On the Vitality of Vitalism', which opens this collection, Monica Greco presents a case for the necessity of the double focus described here. Arguing against the dominant claim that vitalism is obsolete, Greco elucidates its meaning in the work of Georges Canguilhem. Critical of the way in which vitalism is often dismissed on the basis of being over-simplified (for example, in its opposition to mechanism or as a form of meta-physics), Greco suggests that its untenability is in fact the starting point of Canguilhem's re-evaluation. What is significant for Canguilhem is that it keeps being necessary to refute vitalism and so its vitality is in part historical:

The imperative to refute vitalism, in a sense, is superseded by the need to account for its permanent recurrence. The question of vitalism acquires a new dimension – a diachronic dimension – that supplements and subverts each one of its settlements.

Vitalism remains vital partly because of its epistemological role within the history of the life sciences. It is 'the negative term of reference against which biological thought and techniques have progressed' and it represents a 'significant motor force' in the history of biology – 'an error but one endowed with 'a positive, perhaps even necessary function'. Vitalism functions in part as an ongoing form of resistance to reductionism 'and to the temptation of premature satisfaction', closure, denial or ignorance. It is significant to Greco's argument that for Canguilhem vitalism should be regarded as 'an imperative rather than a method and more of an ethical system, perhaps, than a theory'. The key to the vitality of vitalism, as Greco sees it, however, relates as much to its ontological as its epistemological role. Its address, as represented in Canguilhem's work, is to the object of life itself and where it may fail to offer a valid representation of life, vitalism serves as 'a valid representative' – a symptom of the specificity of life. As such, vitalism is consistent with a non-essentialist ontology, an ontology 'of what is perma-nently suspended between being and non-being'.

With these epistemological and ontological resignifications of vitalism established, Greco turns to her chosen homologues in systems and complex-ity theory, finding in the latter a concept which functions in similarly provocative and performative ways epistemologically and which is usually recognized as almost constitutive of the "living object"'. Importantly, she argues that complexity can be seen to constitute the kind of ethical impera-tive that Canguilhem identified with vitalism – the imperative 'that we acknowledge a sensitivity of the world to our interest in it, and to the forms in which this interest is expressed'. In these ways, Greco's conclusions about the ethical imperatives afforded by vitalism and complexity provide one of the starting points of this collection.

But of course the historical and philosophical inspiration for thinking (about) vital processes in this introduction and the following chapters comes from a variety of sources, not just Canguilhem. These include Bergson's notion of duration (1988), in which there is movement but not *a* mobile – things change, but change is not external to the object nor imposed upon it. We also draw on Alfred North Whitehead's concept of concrescence (which might be likened to becoming) in which '[t]here are not "the concrescence" *and* "the novel thing": when we analyse the novel thing we find nothing but the concrescence' (Whitehead, 1985: 211). Other sources of inspiration include Deleuze's (1993) notion of the fold (enfolding and outfolding spatial and temporal differentiations) and Donna Haraway's 'dirty ontology' (Keller, 2002: 62). While we do not under-estimate the very real differences between these understandings, we believe they share a number of assumptions, including – perhaps above all else – a view that entities are constituted in relations. Process, in other words, is characterized by a radical relationality: the (social and natural) world is understood in terms of constantly shifting relations between open-ended objects. This is not to suggest that there are relations *between* pre-existing entities or objects. Instead, objects, subjects, concepts are composed of nothing more or less than relations, reciprocal enfoldings gathered together in temporary and contingent unities. Furthermore, since a relation cannot exist in isolation, all entities can be understood in relation to one another.

The significance of relationality in process thinking in this collection is that it acts as a 'lure for life', an enticement to move beyond the conflation of life with the (life) sciences, to conceive life as not confined to living organisms, but as movement, a radical becoming. In process thinking, relations and relationality cut through and across all spheres, regardless of the distinctions that are drawn between them (between the cultural, the natural, and the artificial, for example). Indeed, distinctions are themselves an aspect (an effect) of the differentiation of processes. In this way of thinking, it is not enough to see nature as an effect of culture or vice versa. The existence of changing distinctions, of ontological differentiations, between nature and culture is instead an effect of particular relations. Indeed, this is another factor in the contemporary appeal of process thinking. It is a response to the impasse that characterizes the conduct and critique of scientific materialism, in, for example, the natural sciences, psychology, and economics. On the one hand, scientific materialism continues to gain ground, regardless of the disasters (ecological, 'artificial', ethical) that pursue it. On the other, neither 'culture' nor the 'subject' (or the subject as it is personified in the artist or scientist, a company, a class or a nation, in language or in discourse) has proved an adequate starting point for enquiry or critique. Thinking in terms of process, in contradistinction to both of these positions, recognizes the *relations* between the traditions that privilege culture, on the one hand, or nature, on the other – their mutual 'ecology', as Isabelle Stengers might call it (1999: 202) – and attempts to displace the bifurcations that they install.

A further advantage of process thinking, as we understand it here, is that the co-ordinates of space and time are not understood to be external to (relations between) entities. Change, that is, does not occur *in* time and space. Instead, time and space change according to the specificity of an event. The event makes the difference: not in space and time, but to space and time. Importantly, motion and change are attributable to differences *within* an event. Duration is the field for the event: there are as many dura-tions as there are events. The accent on process thus loosely corresponds to a privileging of an ontology of *becoming over being* – to a flux that cannot be understood except in terms of process, or passage. This privileging of becoming over being is shared with recent work on performativity (Butler, 1993) in which the on-going temporality of the matter of bodies is acknowl-edged. However, the focus on relationality in process thinking represents a departure from an understanding of material processes via language and discourse. Instead, process thinking seeks to acknowledge the *situated heterogeneity* of stubborn facts – of actual things, concrete events, enduring entities – given neither in matter nor in discourse. This, we suggest, is a crucial way in which both sociological and biological reductionism may be avoided. No fact – whether produced in language or designed in the labora-tory – may be taken to be a privileged example of a stubborn fact (Stengers, 2002: 250).

The attention to stubborn facts also ensures that process thinking is not only concerned with the ephemeral or the liberatory (as is sometimes suggested of theories of performativity and becoming). A stubborn fact is something that cannot be undone. Although the route – the path, the tradition, the inheritance – that has led to the 'becoming' of a fact is contin-gent (the result of 'invention, imagination, and passion' [Stengers, 2000: 251]), it is also irreversible. This, indeed, is one of the reasons that no fact is *less* factual than any other fact. Instead, each and every fact is specific to its own becoming. For it is 'how an actual entity *becomes*', Whitehead writes, that 'constitutes *what* that actual entity *is*' (1985: 23). Theories of performativity opened up the possibility of exploring how the 'what' becomes, but somewhat problematically now often stand in the way of more specific analyses of becoming. Process thinking insists we equally attend to the specificity of heterogeneous becomings (of which performativity is just one instance), whether, for example, those of experimental facts or those of evolution. It is an invitation, in other words, to descend into the hetero-geneous, mixed-up world-making activities of Haraway's dirty ontology.

At the same time that facts are irreversible, they are able to be undone; that is, they are not closed. They always give rise to – they are always enfolded in – novel interpretations (new facts). Every time a 'rupture' is identified, a new series of relations are established. Each fact or entity might thus be said to be the resource or potential out of which new entities emerge. A further implication here, then, is that critiques of the social world that come out of process thinking cannot be conceived of solely in terms of 'taking things apart' (deconstruction), for every taking apart is at the same time a *reconstruction* of relations and relationality. But this reconstruction may be done in different ways, in ways that may be more or less inventive. Thus, some of the articles in this special issue place their emphasis on the conditions under which novel entities are created. Others focus on the limitations, the potentiality that must necessarily be excluded in order that concrete entities emerge out of process. What they have in common is a desire to explain concrete actualities *without explaining them away*. For the singular relations that compose a stubborn fact follow no general laws, nor can they be understood with reference to cause and effect. There are no general or abstract principles that can explain an event. Instead, abstractions themselves require investigation. It is in this sense that all the pieces in this special issue can be said to be empirical. They investigate the particularity (the singularity) of their 'objects'. In doing so, moreover – in exploring and seeking to understand an event – these pieces also *intervene* in the event. Rather than merely realizing what is possible, therefore, this collection aims to contribute to the shaping of vital processes, of processes of actualization, the actualization of potentiality.

Our specific concern in this special issue is to apply the historical function of vital thought to disturb and unsettle recent developments in the fields of (bio)science and technology. These developments often put social scientific analytic tools and conceptual categories at risk by challenging some of the abstractions on which they are conventionally based. For instance, in an era of 'designer' pharmaceuticals, of genetically modified and patented organisms, and of computer synthesis, the distinctions between nature, culture, and artificiality no longer hold fast (we might well ask, of course, if they ever did). At the same time, relations between information, knowledge and science are open to contestation, and, as invention (rather than discovery) is often celebrated as the *modus operandi* in bioscience and medicine, the notion of creativity loses its exclusive association with art and aesthetics. Insofar as process thinking *cuts across* genres, species, and disciplines with little regard for the distinctions that are identified between them, it can be said to be symptomatic of these developments. It is no less important for that, however.

As a term that is relevant but not exclusive to the natural sciences, process offers one way to engage with bioscience and technology without necessarily validating the privileged position claimed by (and attributed to) them, particularly in relation to 'life'. Indeed, insofar as the notion of process is potentially able to extend 'life' beyond the narrow boundaries defined as legitimate by the natural sciences, it indicates not only that the natural sciences do not exist in *epistemological* isolation, but also that their objects do not exist in *ontological* isolation either. In relation to the first of these, process thinking contributes to the turning of an opposition into a contrast. This is what Isabelle Stengers, for example, aims to do: to transform scientific claims into more concrete ones, and thereby to deprive them of the 'science versus opinion' dichotomy that is 'both the condition for science and its confirming result' (Stengers, 2002: 249). But Stengers' cosmopolitics also suggests that the human and life sciences share something in common ontologically:

The pre-fix 'cosmo' takes into account that the word *common* should not be restricted to our fellow humans, as politics since Plato has implied, but should entertain the problematic togetherness of the many concrete, heterogeneous, enduring shapes of value that compose actuality, thus including beings as disparate as 'neutrinos' (a part of the physicist's reality) and ancestors (a part of reality for those whose traditions have taught them to communicate with the dead). (Stengers, 2002: 248, references omitted)

Recent developments in the biosciences and technology certainly seem to be an open invitation to attend to the concrete ways in which science produces not 'just' facts but also, more dramatically, values, modes of exist-ence, or kinds of life. Consider, for example, Prozac, Viagra and other so-called 'lifestyle' drugs, which raise challenging questions about ways of living, about the relations between enablement and enhancement, and between the natural and the normal (see, for example, Marshall and Katz, 2002; Parens, 1998). Or the use of trademark law to protect exclusive ownership of life forms, such as the Oncomouse™ (Haraway, 1997). However, it is precisely in this context – a context where entities once considered to be almost wholly confined to the sphere of the biological and/or natural (and which were also often, by the same token, deemed to be fixed and static) are being recognizably shaped by social and cultural values – that the combined attention to the 'vital' as well as to processes is especially crucial. For it is *not* the case, we would argue, that the 'opening up' of the vital order to cultural, political, and moral interventions renders the opportunities and possibilities of that order *reducible* to such inter-ventions. On the contrary, as Monica Greco argues, 'The point about a vital order is not that it is an order of existence uncontaminated by human artifice or by the "social"; the point is that it involves the social *and much more besides*' (2004). The challenge that the notion of the vital – in whatever form it takes – poses then, is a challenge to all kinds of reductionism: socio-logical and cultural no less than biological.

If the article by Greco provides an introduction to the historical function of vitalism, that which follows – by Suhail Malik – is an example of the contemporary value of the challenge that vital thought provides. 'Information and Knowledge' is a sustained critique of reductionism, specifically, of the idea that information is a reduction or instrumentaliza-tion of knowledge. In Lyotard's notion of the postmodern condition, Malik identifies the suggestion of a move or transformation 'from knowledge, which has a meaning and value that is intrinsic to it, to information, which has an extrinsically determined meaning and value', an exchange-value rather than a use-value and a reductive operational ontology. In order to contest this account of transformation and to establish the extent to which information is necessarily an instrumental category, Malik opens out the term and explores its uses. He finds an inconsistent but not incoherent category typi-cally marked by abstract characterizations such as signal and instruction but in need of further exploration in order to establish what those signals and instructions occasion: in other words, 'what something like "instrumentality" actually amounts to' within particular systems. Drawing on arguments from systems theory, biology and social theory, Malik addresses the role of information in the development of all orders of living and social systems from the pre-cellular organic to post-industrial level. He proposes that it is a 'situated event', both dependent on and generative of meaning and memory. Contrary to the received criticism of information as instrumentalization, he finds that there is no reduction involved in the use of information but instead a complexification of meaning at the centres of human life.

Drawing on developmental systems theory, Malik shows that what counts as information in a living system is contingent on the system as a whole at a given time. Information is 'a difference that makes a difference'. It is a relational concept which presupposes structure – not itself a structure but 'rather an event that actualizes the use of structures'. Information is an event because it makes a difference or alters the state of the system, creating (new) meaning. The term 'eventive' is introduced by Malik as a way of indicating the dynamic operation of information and how 'the each time event of information is transitive, verb-like and adjectival rather than substantive and noun-like'. The term is also meant to parallel 'invention' and therefore to encompass the anthropo-technical as well as biological scale of human development: the 'epiphylogenetic' complexification of meaning that removes information from, for example, the reductive characterizations of an immutable genetic code. In this way, he introduces a further theme of the collection: how to think the human. Drawing in particular on the work of Stiegler, and arguing against that of Virilio, Malik's point is that information is not external to the interests of the human: 'It is rather another name for its continued development, and the development of [its] meaning, as a complexly constituted phylum.' Instrumentality is reclaimed as (human) evention.

Through a case study of pharmaceutical research and development ('Pharmaceutical Matters: The Invention of Informed Materials'), Andrew Barry further contributes to the study of invention. He does so by challenging a solely technical image of chemistry as well as received ideas about the static self-identity of molecules, as established, for example, in the periodic table. Adopting and developing Bensaude-Vincent and Stengers' concept of informed materials, which displaces the binarism of information and materiality, Barry places invention (rather than innovation) at the heart of chemistry. He argues that chemical R&D is not merely innovative but is 'itself creative or inventive' in that it invents informed materials and, indeed, new methods for the invention of informed materials.

Drawing on both Gabriel Tarde's account of invention and Alfred N. Whitehead's 'metaphysics of association' (where social refers to the association between non-human as well as human entities), Barry argues that invention is process-based rather than progressive, it is contingent, 'irreversible and path-dependent' but also 'indebted to romantic notions of individual creativity'. In Whitehead's theory of the organism, Barry finds an appropriate historical and social framework for the invention of novel entities. This framework is centred on the notion of the molecule as an event which endures, but which does not necessarily remain the same. Citing Whitehead, Barry suggests that 'a molecule is a historic route of actual occasions; and such a route is an event'. Motion and change must be ascribed, in other words, to the differences between the successive actual occasions that are contained within an event. Such actual occasions or entities, moreover, are not bounded 'but are extended into other entities, while folding elements of other entities inside them'. As such, they are insufficiently 'bare' to be 'discovered'; rather, chemical 'invention' is concerned with the proliferation of the forms of existence of molecules.

Focusing his analysis on the study of a company (ArQule) directed towards the development of chemical entities, Barry demonstrates that what pharmaceutical companies actually do is invent novel entities. The concept of informed materials, especially insofar as it captures relationality, illustrates the particular character of this novelty. Informed materials are not those which are shaped or formed externally, as one might impose shape on a mass of material, but those which are developed to become 'richer and richer in information'. In the particular case studied by Barry, the invention of informed materials depends on 'a detailed comprehension of the microscopic structure of materials', which can then be manipulated. But invention is an aspect not only of something known to chemists as 'chemical space'. — 'a relational space, the co-ordinates of which are governed by the particular chemical process under investigation' — but also of intellectual property laws. The molecules produced in pharmaceutical laboratories 'are rich in information about their (global) legal and economic, as well as their chemical relations to other molecules'. Thus, Barry argues that molecules should be viewed not as discrete objects, but as constituted in their relations to complex informational and material environments and that their endurance should be understood to depend 'upon the multiplication of different forms of informed material'.

In 'The Performativity of Code: Software and Cultures of Circulation', Adrian Mackenzie looks at the way in which computer code gains significance or becomes operational through complex processes of circulation that cut through and across, and which contingently reconstruct, the relationality of 'the social' and 'the technical'. Through a case study of Linux which is highly attendant to the historical, architectural and contextual specificity of software or code objects, Mackenzie offers an analysis of this particular technical 'culture-object' which goes far beyond the formal properties of the object itself. The case study instead considers how the force or performativity of Linux 'can be understood more [clearly] as the stabilized nexus of diverse social practices, rules and personae'.

Mackenzie argues that the technical performativity of objects such as Linux go beyond speech-based notions of code as instruction to their various practices and processes of mediation and circulation. In this way,

Mackenzie seeks to extend existing concepts of performativity and to account for processes 'in which the creation of meaning is not central, and in which processes of circulation play a primary role'. Given 'the many ways, sites and levels at which performativity works, and the large-scale differences of class, race, gender and sexuality for which the conceptual applications of performativity have been developed', Mackenzie writes, 'how is something quite thing-like or even infrastructural like Linux at all relevant to the contemporary performativity of power?'. Drawing on Butler and Derrida, Mackenzie emphasizes the idea that performativity is a facet of the extra-linguistic effects of linguistic praxis, the redoubling of utterances by acts. Even though computer code, 'an exemplar of formal clarity and univocity' might not obviously lend itself to a performative analysis, still it cannot exist or operate 'apart from a prior set of practices which allows it to do things'. Detailing the various forms in which Linux circulates, and the way in which it succeeds, partly by 'covering over' the gendered and other authorizing contexts which give it force, Mackenzie gives an account of the collective performativity of code, its ability to objectify linguistic praxis or 'enact something' through processes of circulation. It is through these processes of circulation, he suggests, that Linux endures.

Celia Lury's article, ' " Contemplating a Self-portrait as a Pharmacist": A Trade Mark Style of Doing Art and Science', further explores the themes of creativity, event and invention in a discussion of the work of the artist Damien Hirst and, more particularly, through the proposition that the name Damien Hirst functions as a brand. No longer the mark of an originary relationship between the producer and the product, the brand name, Lury argues, is increasingly 'the mark of the organization of a set of relations between products in time'. In view of this alternative understanding, the article addresses how the brand, 'as a response to the emergence of the world we live in as media, transforms "the conditions of the emergence of novelty"'. Once again, as in the article by Mackenzie, the emphasis is on the significance of the centrality of processes of circulation, of meta- or mixed-media, for how things come into being, persist, and have effects. In particular, she seeks to explore how the emergence of the relational object of the brand contributes to the transformation of the author-function in ways that challenge and enervate contemporary conceptions of what constitutes 'science' and 'art'.

Through an analysis of the spot paintings, Lury demonstrates the absence of authorship in Hirst's work; his renunciation both of (himself as) origin, and of originality. She shows how, instead, Hirst uses a (his) name to structure and organize the relations between things. The meaning created is, characteristically, movement, immediacy and open-endedness: the Hirst brand may 'be characterized as a staging of a lived relationship' – as (in the title of one exhibition) *Life/Live*. Further examples of the distinctiveness of the Hirst brand range from the cow in formaldehyde to the staged life-cycles of flies and butterflies, and an edition of prints that allude to drug packaging (just as the spot paintings themselves were often named after drugs).

Crucially, Lury argues that this brand of work constitutes more than a commentary *on* science: that it is itself the staging of a kind of science or, more specifically, of a pharmaceutical kind of invention, creativity or 'discovery'. But while she argues that Hirst's framing of the relations between entities can be considered experimental, she simultaneously maintains that the practices of pharmacy may 'increasingly be understood as science conducted under the sign of the brand name'. It is the brand, then, that 'provides the experimental conditions for "the emergence of novelty"', and trade mark style which offers a new way of linking art and science.

Performativity and authorship are also central to Lisa Adkins' examination of the reworking of relations between people and property in the new economy. In 'The New Economy, Property and Personhood', Adkins suggests that analysts of the new economy have side-stepped the issue of 'how personhood itself may be materially reconstituting in the new economy': 'when people are discussed they are assumed to be largely in control of and, indeed, to own their own identities and bodies'. Following the work of Pateman, Adkins takes issue with the assumption of a 'social contract model' in which people may own property in their own personhood and then disentangle the attributes or qualities of that property in order to trade them. What Adkins suggests, in contrast, is that in the new economy, 'people cannot unproblematically claim to own and straightforwardly accumulate property in the person, since the relations between property and people are being restructured'.

Adkins draws her inspiration from work at the boundary of feminism and science and technology studies – principally that of Haraway and Callon. This work is concerned with the epistemological and ontological questions wrought not only by the commodification of social relations, but by the commodification of life itself (for example, in the patenting of cloned transgenic organisms) and by the move from (natural) kind to brand – and indeed from brand to kind. The shifting ground of 'natureculture' identified by Haraway, the ongoing processes of denaturalization and renaturalization, are broadly analogous for Adkins with those of attribution and reattribution, of qualification and re-qualification, in the new economy. In these processes, qualities previously associated with people are not owned in the conventional sense (as forms of property inherent in the person) but are rendered detachable. As she puts it, 'in the new economy characterized by data, knowledge and service intensity, gender is detachable from the person and indeed may be alienable as a form of property'. This is indicated in corporate demands for flexibility and adaptability, and by workplace exercises which seek to 'scramble' male and female characteristics in order to optimize efficiency. Detached, mobile, alienable, in short, performative, gender is figured as a workplace resource, albeit one whose ownership is complicated by the significance awarded to 'audience' effects and attempts to standardize or 'performat' these. The significance of this gender performativity has been recognized but underplayed, Adkins argues, as a result of the hegemony of the social contract model, the displacement of which enables her to provoke further consideration of notions of personhood in the new economy.

In 'Computing the Human', N. Katherine Hayles focuses our attention on radical relationality by exploring the feedback loop between humans and intelligent machines, and future predictions about the novel converged entities which constantly cycle through present thinking (sensing and acting). Her premise is that 'visions of the future, especially in technological advanced eras, can dramatically affect present developments' and this should direct our attention to works which project a future in which humans and intelligent machines become virtually indistinguishable. Rather than condone (or condemn) this kind of speculation, Hayles seeks to explore the influence this future thinking has on present concepts, since what is at stake is being able to contend 'for how we now understand human thinking, acting and sensing – in short, how we understand what it means to be human'.

Her piece centres on the Sense–Think–Act paradigm used in AI because it 'defines the necessary behaviours an entity needs to interact with the world'. This becomes the basis for convergence scenarios, which, importantly, are tied into narratives of progress and evolution: 'The pressure to see *Homo sapiens* and *Robo sapiens* as essentially the same emerges as a narrative of progress that sees this convergence as the endpoint of human evolution.' While she remains equivocal about evolutionary endpoints (who could be otherwise?), Hayles is unequivocal about the performativity of all narratives that take intelligent machines as a reference point for descriptions of the human. Through her analysis of Acting, Sensing and Thinking and of the kinds of entities which characterize and are characterized by this central paradigm, she gives consideration to technical and other specificities including the role of biological and evolutionary computing. If evolution does not describe a teleology with the convergence of humans and intelligent machines as its endpoint, then what kinds of relationality does it construct or make possible? For Hayles, there are complex relationalities between humans and intelligent machines that do not resolve neatly into sameness or difference but that do seem to allow for creativity and invention. For example, she finds that a serious case can be made for (Koza's) genetic programs possessing 'the human attributes of creativity and inventiveness'. Although this argument cannot be said to support the future idea of convergence or sameness, it is deliberately counterposed to arguments, such as those of Francis Fukuyama, who emphasize difference and seek to reinstate notions of human nature which might be said to be nostalgic, rooted in the past. Perhaps it can be said, then, that Hayles is not attributing human/machine agency to evolution (either as endpoint or process) but rather sees agency, including the power to determine the future, as a facet of human/machine co-evolution.

Staying with the theme of evolution, Sarah Kember ('Metamorphoses: The Myth of Evolutionary Possibility') offers a critique of the role of genetic or what she calls abstract evolution across the increasingly eroded

boundaries of art and science in contemporary culture. Focusing on artificial life and transgenesis, where the possibilities and processes of radical
human/animal/machine relationality and the emergence of novel entities are
highly visible, she argues that abstract evolutionism, far from facilitating
such processes, actually closes them down. Drawing on François Jacob's
critique of modern or neo-Darwinism, she elaborates the myth of evolutionary possibility where the evolution in question is abstract, informationalized (located at the level of genes rather than organisms) and
over-extended – not so much a hermeneutic as a 'sterile belief'. Without
dismissing the potential of evolution *per se* or over-looking existing challenges and conflicts coming from both inside and outside the field of evolutionary theory, Kember shows how evolutionary possibility is currently
mobilized through the myth of metamorphoses and is thereby attached to
assumptions about radical change and transformation in the human
condition. The problem here, as Kember demonstrates with reference to the
work of Marina Warner, is that in the original myths of metamorphoses
(Ovid, Dante) transformation is figured either as punishment or perfection,
transgression or transcendence. As such, it often conjures up the perverse
image – the monster, mutant, hybrid – which acts as a salutary reminder of
the sanctity of human self and species identity. Kember seeks to illustrate
the conservatism inherent in the (metamorphic) myth of evolutionary possibility through an examination of two sci-art works representing (though not
of course exhausting) artificial life and transgenesis. These are *Galápagos*
by Karl Sims (an interactive media installation in which users evolve three-
dimensional animated life-forms) and *Genesis* by Eduardo Kac (a transgenic
artwork in which Kac challenges the word of God through a collective interactive process of Darwinian evolution-as-reinscription).

In both works, so Kember suggests, abstract evolutionary possibility
is reified over and above any kind of interactivity, connectivity or relationality between humans and computers or between species. Indeed, it is
reified over and above the actual material circumstances of the emergent
life-forms themselves and as such lacks what Jacob refers to as a dialogue
between the possible and the actual. Evolution produces no-thing new and
all that metamorphosis is metamorphosis itself. Kember's purpose here is
not only to expose what she sees as a certain imaginative poverty but also
to draw attention to the problematic relation between evolution and
becoming in contemporary discourses and practices. Evolution, in Kember's
account, is relatively impotent and 'shadows a quest for metamorphoses as
"becoming versus Being in its classical modes"'.

Drawing together our thematic concerns with becoming, events and
life (itself), Mariam Fraser ('Making Music Matter') examines a performance
piece by Bruce Gilchrist, *Thought Conductor # 2* (*TC2*), in which the signals
generated by a person linked to an EEG are converted into musical notation
and played, live on stage, by a string quartet. In an echo of one of the questions asked of the original inspiration for Gilchrist's piece, John Cage's
4'33", which has been described as one of the first happenings in America,

Fraser asks of *TC2*: what *kind* of happening is this? She answers this question through a comparative analysis of the relations between author/composer, score and sound in both *4'33"* and *TC2*, and of the different concepts of presence and life that each work evokes.

Cage's visit to an anechoic chamber – in which he expected silence but heard, instead, the sound of his own nervous system and blood circulating – led him to revise his understanding of both silence (as 'the absence of intended sounds') and music (as 'an attentiveness to the sheer immediacy of an absolutely contingent conjunction of incidental sound'). These understandings inform *4'33"*: 'Composed according to chance principles, the three movements of silence . . . leave plenty of space for indeterminacy.' Constituted by 'accidents of performance' not directly connected to the score or notation, this work challenges musical conventions and, in particular, the laws of determinacy that govern the relation between author/composer and text/score, and between score and sound. A means of accessing ready-made, environmental, or unintentional sounds which happen *in* the here and now, *4'33"* introduces the listener to the aliveness, the presence of life, in a specific space and time. Although Gilchrist's *TC2* is equally alert to contingency and indeterminacy, Fraser argues that it does not assign these qualities to, or seek to discover them within, anything independent of the performance at all, including or especially the 'life' or even the brain waves of the individual linked to the EEG. Rather than directing the musician towards the capture of sounds that exist *in* space and time, the score of *TC2* is instead 'the product *of* the time of the performance during which sounds are created'. The relation between score and sound is not reversed here then, but is rather inextricably enfolded, as 'elements in a pattern'. Fraser, drawing on Whitehead's theory of the organism, re-conceives liveness or presence as pattern (out of which time and space emerge) and performance as event (defined not in terms of the relations between individual entities or components but as a singular 'becoming-together', which is creativity). These re-conceptions enable her to argue 'that *TC2* does not introduce the listener *to* life, but that it is *itself* alive'.

References

Bergson, H. (1911) *Creative Evolution*. Trans. A. Mitchell. London: Macmillan.

Bergson, H. (1988) *Matter and Memory*. Trans. N.M. Paul and W.S. Palmer. New York: Zone Books.

Butler, J. (1993) *Gender Trouble*. London and New York: Routledge.

Deleuze, G. (1993) *The Fold: Leibniz and the Baroque*. Trans. T. Conley. London: Athlone.

Greco, M. (2004) 'The Politics of Indeterminacy and the Right to Health', *Theory, Culture & Society* 21(6): 1–22.

Haraway, D. (1997) *ModestWitness@Second_Millenium. FemaleMan_Meets_OncoMouseTM*. New York and London: Routledge.

Keller, C. (2002) 'Process and Chaosmos: The Whiteheadian Fold in the Discourse of Difference', in C. Keller and A. Daniell (eds) *Process and Difference: Between*

Cosmological and Poststructuralist Postmodernisms. Albany, NY: State University of New York Press.

Marshall, B. and Katz, S. (2002) 'Forever Functional: Sexual Fitness and the Ageing Male Body', *Body & Society* 8(4): 43–70.

Parens, E. (1998) ' " Is Better Always Good?" ' The Enhancement Project', *Hastings Center Report* special supplement, January–February: S1–S17.

Stengers, I. (1999) 'Whitehead and the Laws of Nature', *SaThZ* 3: 193–206.

Stengers, I. (2002) 'Beyond Conversation: The Risks of Peace', in C. Keller and A. Daniell (eds) *Process and Difference: Between Cosmological and Poststructuralist Postmodernisms*. Albany, NY: State University of New York Press.

Whitehead, A.N. (1985) *Process and Reality*. New York: The Free Press.

On the Vitality of Vitalism

Monica Greco

MANY OF those who share an interest in the life sciences today, perhaps most, would agree with the claim that vitalism is obsolete. Some prominent biologists now use 'vitalism' as a derogatory label associated with lack of intellectual rigour, anti-scientific attitudes, and superstition (see, e.g., Dawkins, 1988). Other scientific commentators treat the term more seriously, but equally arrive at the conclusion that vitalism is an untenable perspective. Prigogine and Stengers (1984), for example, have described vitalist concepts as meaningful for biology within the broader scientific context characterized by Newtonian physics, but as having been made redundant by 20th-century developments both in physics and in (molecular) biology.

The claim that vitalism is obsolete – the ease with which this proposition is apparently accepted – might serve as a useful way of approaching the particular status of this concept in the work of Georges Canguilhem. The link is pertinent at different levels, in different ways. First of all, there is the point that Canguilhem addresses vitalism precisely in so far as it is constantly refuted, constantly declared 'obsolete' – a point to which I shall return in more detail below. Second, there is a question of the extent to which Canguilhem's own propositions may be 'obsolete', on account of their vitalism.

This second question seems to be a concern in several commentaries that offer positive appreciations of Canguilhem's life and work, but that also seem to share an intent to qualify and circumscribe the vitalist aspect of his thought. Dominique Lecourt, for example, exposes the 'misadventures' of Canguilhem's vitalism in so far as it relies on a notion of biological individuality that has been exploded by the development of molecular biology since the 1950s. According to Lecourt, while Canguilhem rightly highlighted the need to understand and preserve vitalism as a meaningful intellectual and ethical demand rather than a positive philosophy of life, this did not prevent him from attaching a substantialist and anthropocentric ontology to the term. For this reason, Lecourt suggests that 'perhaps we should abandon the word "vitalism", given that it is so ambiguous' (1998: 223). In the same collection, Nikolas Rose points to the languages of 'genetic evolutionism, molecular biology, biomedicine and biotechnology, their new institutional sites . . . and their new techniques' as new formulations of the problem of life that question fundamentally the organic image that underlies the notion of a vital normativity described by Canguilhem. It is clear, Rose writes, 'that normativity no longer can be understood in terms of the self-regulation of a vital order – if it ever was. Normativity now becomes a matter of normality, of social and moral judgments about whether particular lives are worth living' (1998: 165). As a result, Rose too proposes that 'the productivity of Canguilhem's reflections on norms in life lies less in his insistence on the vitality of life than on the light that it sheds on the character of those other norms that traverse our culture' (1998: 164).

Judging by these texts, Canguilhem's vitalism can and should be isolated as a mere aspect of his thought, and not necessarily the most relevant or defensible. The suggestion is that we should read his work despite, and not because of, its vitalism. It is against this background, and against this closure, that I believe it is necessary to pose anew some basic questions of clarification. These basic questions include: What sort of vitalism is Canguilhem's vitalism? Is it justified to suggest that this form of vitalism is obsolete? What is the ambiguity of 'vitalism', and should the term be abandoned on its account?

I The Vitality of Vitalism: Two Meanings

The term 'vitalism' is most readily associated with a series of debates among 18th- and 19th-century biologists, and broadly with the claim that the explanation of living phenomena is not compatible with, or is not exhausted by, the principles of basic sciences like physics and chemistry (see Benton, 1974; Lenoir, 1982; Cimino, 1993). However, scientists and philosophers have continued to address vitalism – if mostly in order to reject it – well into the second half of the 20th century, in connection with classic concepts such as mechanism and reductionism, but also in connection with the concepts of emergence, complexity, artificial intelligence, and with approaches such as information theory and cybernetics (see Carlo, 1966; Hein, 1968a, 1968b, 1969; Ackermann, 1969; Bronowski, 1970; Hoyningen-Huene and Wuketits, 1989; Rapaport, 1995; Emmeche et al., 1997).

Not many authors acknowledge the semantic polyvalence of vitalism, and it is commonly assumed that vitalism necessarily involves some reference to metaphysical principles, some degree of teleological thinking, and the opposition to mechanism. While these features do apply to some vitalists, Benton (1974) has shown that they represent a gross simplification and misrepresentation of the variety of meanings associated with the term.[1] For our purposes it will suffice to refer to the broad distinction proposed by Wuketits (1989) between 'animist' and 'naturalistic' varieties of vitalism, where the first is explicitly metaphysical and teleological in orientation, while the second posits organic natural laws that transgress the range of physical explanations. Both varieties of vitalism are described by Wuketits – who speaks from the perspective of general systems theory – as 'untenable in the light of modern biological research'.[2]

To understand the specific form of Canguilhem's vitalism, we need not ignore this proposition of untenability. On the contrary, as I have already suggested, it is necessary to take it as a point of departure. Canguilhem proposes that we treat as significant the historical 'vitality of vitalism' – the fact that the imperative to refute vitalism has had to be continually reiterated up until the present. In so doing, he is also proposing that we revise the notion that what appears to be methodologically untenable from the perspective of contemporary science may be ignored or dismissed as obsolete. The imperative to refute vitalism, in a sense, is superseded by the need to account for its permanent recurrence. The question of vitalism acquires a new dimension – a diachronic dimension – that supplements and subverts each one of its settlements.

The theme of the 'vitality of vitalism', therefore, can be read in the first instance as an answer to the epistemological problem of assessing the value of episodes in the history of science, when this problem is posed specifically in the context of the life sciences.[3] As the negative term of reference against which biological thought and techniques have progressed, vitalism represents a significant motor force in the history of biology (Canguilhem, 1975: 84). It is the trope through which, and in answer to which, the life sciences have come to constitute their own specific domain distinct from those of physics and chemistry. It is an error, but an error endowed with a positive, perhaps even necessary, function. As Michel Foucault put it in his introduction to *The Normal and the Pathological*:

> if the 'scientificization' process is done by bringing to light physical and chemical mechanisms . . . it has on the other hand, been able to develop only insofar as the problem of the specificity of life and of the threshold it marks among all natural beings was continually thrown back as a challenge. This does not mean that 'vitalism' . . . is true. . . . It simply means that it has had and undoubtedly still has an essential role as an 'indicator' in the history of biology. (1989: 18)

These considerations account for a relationship that Canguilhem establishes with vitalism as an element in the history of science. Canguilhem endorses vitalism conditionally, in that the historicity of vitalism allows for a resignification of its value for the science of biology. In itself, however, this says little about the way in which we might qualify Canguilhem's own thought as 'vitalist'. For this we must focus on a further resignification, implicit in the first one, but this time concerning what questions we treat as relevant in relation to vitalism.

In *La Connaissance de la vie*, Canguilhem proposed that vitalism should be regarded as 'an imperative rather than a method and more of an

ethical system, perhaps, than a theory' (1994: 288). Like the first one, this
is a resignification that can only occur by evaluating vitalism diachronically,
as a key element in an ongoing dialectical process. What is relevant about
vitalist theories and concepts for Canguilhem is not what they *say* – and
whether what they say is ultimately scientifically defensible – but rather
what they *do*, by providing a form of resistance or antithesis to the recur-
rent possibility of reduction, and to the temptation of premature satisfaction.
Canguilhem, who was in many ways critical of Bergson, would have agreed
with his point that 'the "vital principle" might indeed not explain much, but
it is at least a sort of label affixed to our ignorance, so as to remind us of
this occasionally, while mechanism invites us to ignore that ignorance'
(Bergson, 1911: 42).

Canguilhem is himself a vitalist to the extent that he invites us to
recognize in the form taken by the history of the life sciences the characteristic trace, the specific response, of *life* as their object: '. . . transposing
the dialectical process of thought [that alternates between vitalism and
mechanism] into the real, we may say that it is the object of study itself,
life, that is the dialectical essence; and that thought must espouse its structure' (1975: 85). Thus, while vitalist theories remain scientifically inadequate and philosophically naive, they are nevertheless directly *relevant* to
the problem of life.[4] The oscillation that characterizes biological thought,
of which the alternative between vitalism and mechanism is but one expression, is the symptom of a form of knowledge marked by a paradox: the
science of life is, itself, a manifestation of the activity of the living, a manifestation of its own subject matter. Once it is understood performatively, as
resistance and excess with respect to the remit of positive knowledge,
vitalism therefore appears valid – not in the sense of a valid representation
of life, but in the sense of a valid *representative*. In other words, and to
reiterate: it is not as an account of life that vitalism appears viable; rather,
it is as a symptom of the specificity of life that its recurrence should be understood. To erase the contradiction that vitalism provides, to dismiss it as a
weakness of thought, is to silence life, and to become ignorant of ignorance.
The 'vitality of vitalism' thus can be read in a second sense as pointing,
beyond a question of epistemology, to an ontology. It points not only to a
dialectic movement internal to knowledge, but also to the movement that
links knowledge with its condition of possibility, life. 'To understand the
vitality of vitalism,' writes Canguilhem, 'means to engage in a search for the
meaning of the relationship between life and science in general' (1975: 85).
His vitalism is an assertion of the originality of life, understood also, and
especially, in the sense of its logical priority with respect to knowledge:
'once we recognize the originality of life, we must "comprehend" matter
within life and the science of matter, which is science *tout court*, within the
activity of the living' (1975: 95). It is impossible, therefore, to confuse
Canguilhem's position with the forms of what he calls 'classical vitalism' –
whether these be animist or naturalistic, to return to the broad terms of
Wuketit's typology. As Canguilhem explains, classical vitalism commits a

philosophically inexcusable mistake when it takes the 'originality' of life to mean that life constitutes an 'exception' to the laws of the physical *milieu*. Classical vitalism, in this sense, is a purely reactive form of thought: it implicitly acknowledges the logical priority, and the normativity, of the world described by the sciences of physics and chemistry. The originality of life cannot be claimed for a segment of reality, but only for reality as a whole. Biology must affirm its own 'imperialism' (1975: 95).[5]

To avoid the possibility of misunderstanding, Hertogh has proposed that we should call Canguilhem's vitalism a 'polemical vitalism'. 'This notion of life', Hertogh writes, 'has a critical and no positive meaning' (1987: 123). Like Lecourt, Hertogh seems prepared to accept the vitalism in Canguilhem, but only with the proviso of a strong qualification; the link with ontology in particular must be treated with suspicion. Yet, I want to suggest, it is possible to misunderstand Canguilhem in more than one way; that is, not only in the naive sense of failing to see the difference between his vitalism and classical forms of the idea, but also in the sense of failing to see a certain continuity between them, by underplaying the reference to the ontology of life and underestimating its significance.

II Contexts of Relevance

In the remainder of this article I will approach this aspect of the question by relating Canguilhem's vitalism to accounts of systems theory and complexity, provided by Franz Wuketits and Isabelle Stengers respectively. This (all too brief) comparison is intended as a programmatic invitation to seek a homologue to Canguilhem's vitalism in contemporary forms of scientific thought, with which one might return to the question of whether this vitalism is 'obsolete'. I might begin, therefore, by stressing that when Canguilhem advocates an 'imperialism' of biology, when he invites biologists to shed their modesty in order to universalize their conception of experience, he sees this position as being perfectly compatible with the tenets of contemporary physics. It is through relativity theory and quantum mechanics that, in the first instance, we have been obliged to reconceptualize the terms that link, or indeed define, the subject and the object of knowledge:

> The environment [*milieu*] within which life is seen to appear only has the meaning of environment through the operation of the human living being, who performs measurements . . . that bear an essential relation to technical instruments and procedures. . . . [W]e have come to discover that, in order for there to be an environment, there must be a centre. What gives an environment the meaning of conditions of existence is the position of a living being, referring to the experience it lives in its totality. . . . Thus understood, a biological point of view on the totality of experience appears perfectly true [*honnête*] both to the man of knowledge, especially the physicist, and to man as a living being. (Canguilhem, 1975: 96)

Critics like Dominique Lecourt (1998) have taken issue with this proposition, suggesting that Canguilhem transfers his own concepts into a reading of contemporary scientific developments, instead of letting these instruct his philosophy.[6] For Lecourt, it is impossible to reconcile the primacy accorded to the relationship between terms with the idea that one of those terms, namely the living individual, constitutes an absolute centre of reference at every level' (1998: 220). Lecourt claims that Canguilhem's use of the category of 'living individual' is both anthropomorphic and anthropocentric, and he presents these as forms of self-contradiction and conceptual weakness. Is this actually the case? The question, I should stress, refers not to whether the text actually is anthropomorphic/centric, but to whether this constitutes a conceptual weakness, indeed the sure symptom that a text must be 'obsolete'. I propose that it does not, and that to construe it as such is to miss the point of the epistemological shift that is being proposed.

In support of this view, it is possible to turn to Donna Haraway, who might well be described as one of the most authoritative voices in the contemporary debate on post-humanism. Haraway, like Lecourt, refers to contemporary biology to contest the notion that a stable ontological ground can be provided for the category of the 'individual'. On the basis of an analysis of immunological discourse, she has famously argued that '[a]ny objects or persons can be reasonably thought of in terms of disassembly and reassembly; no "natural" architectures constrain system design' (1991: 212). Equally interesting, but perhaps less often remembered, is how Haraway's argument goes on to develop in the same paragraph. 'Design', she writes, 'is none the less highly constrained. What counts as a "unit", a one, is highly problematic, not a permanent given. Individuality is a strategic defense problem.' What we have here is a notion that there is no *essential* stability or necessity — whether physical or logical — to the category of the individual. But, in the second part of the quotation, we also have the suggestion that categories of the individual, of the 'unit' or of the 'one', may well pertain to a different order of necessity — political, clearly; ethical, perhaps: 'Individuality is a strategic defense problem.'

That this order of necessity might be described as a 'vital' order is indicated by the continued relevance of the notion of the individual in relation to the polarity between health and disease, or indeed life and death. As Haraway writes, in the conceptual vocabulary of the immunologists, '[p]athology results from a conflict of interests between the cellular and organismic units of selection' (1991: 220), or again '[d]isease is a process of misrecognition or transgression of the boundaries of a strategic assemblage called self' (1991: 212). Thus, to return to the point originally made by Lecourt, the fact that the individual, the organism, and indeed the human form, should be regarded as ontologically contingent, does not contradict the perspective (advocated by Canguilhem) that might place the living being at its centre. On the contrary, it is quintessentially an expression of such a perspective, in so far as a vitalist ontology cannot but be an ontology of the contingent, of what is permanently suspended between being and non-being. Canguilhem's anthropocentrism, in this sense, represents the contrary of what anthropocentrism is usually taken to signify by its critics: rather than affirming a right of supremacy, it suggests a kind of humility, an acknowledgment of (inevitable) partiality or, to use Canguilhem's own expression, a form of 'honesty' [*honnêteté*].

It is significant that Lecourt's critique, which is wielded in the name of contemporaneity as a whole, should take molecular biology as its norm. There is no doubt as to the empirical relevance of this norm. As Nikolas Rose has put it, in one respect

> [T]o analyse what counts as true in life, we need to examine who has the power to define that truth, the contemporary role of different authorities of truth, the new epistemological, institutional and technical conditions for the production and circulation of truths and the ethical and political consequences of these truths. (1998: 168)

That the stakes are especially high in relation to molecular biology is, as I have mentioned, obvious in an empirical sense; the discourse of molecular biology is a powerful discourse, a discourse most visibly supported by the resources of power. The question is whether it is possible and desirable to treat this kind of empirical relevance as the norm for relevance *per se*.[7] Or, to turn this into a question: are there other vantage points, in the landscape of contemporary scientific knowledge, that might lead us to regard vitalism as a current form of thought? And can we justifiably ignore them?

One of the possible sources we might turn to, for the purposes of putting Canguilhem's vitalism through the test of contemporaneity, is the development of a systems-theoretical approach to the question of life, along the lines proposed by Paul A. Weiss (1969) and Ludwig von Bertalanffy (1968). Advocates of this approach, such as Franz Wuketits, set it explicitly against both mechanism and vitalism, suggesting that *both* positions can be guilty of relying on a metaphysical understanding of teleology. While this is explicitly the case in vitalist theories, where the specificity of life is addressed in terms of 'vital forces' or principles, mechanism tends to leave implicit the problem of teleology – which means that it remains possible, if inconsistent, to presuppose a metaphysical agency at the origin of life's organization.[8] Evelyn Fox Keller's excellent studies of the productive role of metaphor in the discourse of genetics amply illustrate this point. She explores, for example, how the construction of developmental explanation in genetic research has relied on semantic effects of ambiguity and polysemy in expressions such as 'gene action' and 'genetic programme'. As Keller argues in far more detail than I can reproduce here, each of these formulations allowed for a strategic 'blackboxing' of the problem of teleology at different points in the history of molecular biology (see Keller, 2001; also, less directly, 2000).

In contrast, according to Wuketits, the systems theoretical approach may be described as reinventing mechanism and vitalism by explicitly engaging with the problem of teleological organization, but only after emptying teleology of any reference to metaphysical principles, and to human or superhuman agency. In this way, this approach claims to be able to address the specificity of living beings while meeting the requirements of scientific explanation. Wuketits, among others, suggests that this amounts to nothing less than the emergence of a new paradigm in biology and bio-philosophy. It is a paradigm that involves an 'organism-centred view of life', where organisms are regarded as open, dynamic, homeostatic, and hierar-chically organized systems characterized by fundamental and increasing complexity (Wuketits, 1989: 16–17; see also von Uexküll, 1997). It is a shift, one might add, that appears to parallel the development in the physico-chemical sciences of fields such as far-from-equilibrium thermo-dynamics and chaos theory.[9] Ironically, however, Wuketits proposes the term 'holism' to refer to this 'organism-centred view of life' – despite the fact that holism is itself a term that needs to be purged of metaphysical associations. This is interesting because it points to a difficulty that can no longer appear contingent but is intrinsic to the problem being addressed, and to the terms through which it comes to be addressed. Could systems theory provide the epistemological shift that Canguilhem advocated under the (similarly prob-lematic) term 'vitalism'? And why does it continue to seem necessary to employ these problematic terms, only to endlessly return to the task of having to qualify them?

These questions take us back to a point made earlier concerning the added value that identifying as a 'vitalist' affords besides, or even despite, the descriptive content of the word. In other words, they take us back to Bergson's notion that 'the "vital principle" ... is at least a sort of label affixed to our ignorance'. And I propose that this point may be usefully expanded by relating vitalism to a concept that has come to play an increas-ingly important, if ambiguous, role in contemporary science, particularly within a systems-theoretical framework: the concept of complexity.

In a brilliant analysis, Isabelle Stengers invites us to distinguish the theme of complexity understood as the hallmark of a 'new science', or as the content of the latest scientific descriptions of the world, from the theme of complexity understood as an expression of what might be called a specific *ethos* or discipline of thought:

... if the theme of complexity is potentially *interesting*, and perhaps worthy of surviving compromising usage, it is ... because it rekindles and highlights what is without doubt the most genuinely original aspect of what is called 'modern science'. As Jean-Marc Lévy-Leblond reminds us, the function of scientific thought has less to do with its 'truth' than with its *astringent effects*, the way it stops thought *from just turning in self-satisfying circles*. (Stengers, 1997: 5)

More specifically, the theme of complexity is meaningful in this sense when it intervenes to mark a leap in the order of possible knowledge, and there-fore a difference in the quality of our ignorance, with respect to a specific instance or problem. To clarify this point, Stengers discusses the contrast between the notions of complexity and complication. A phenomenon is *complicated* when the task of predicting its behaviour presents a difficulty due to incomplete information, or to insufficient precision in the formulation of questions, but when in principle it is possible to explain and understand it by *extending* a simple, fundamental model. In so far as the programme of molecular biology is reductionist, for example, it treats the reality of living beings as a tremendously complicated reality, but one that is nevertheless regarded as understandable, in principle, in terms of the model of a chain of physico-chemical determinations. For the situation of *complexity*, on the other hand, Stengers offers the examples of the behaviour of 'strange attractors' and of unstable dynamic systems. A complex situation is one where

> the difficulty of an operation of passage [from the simple to the complex] may not be due to a lack of knowledge, an incomplete formulation of a problem, or the enormous complication of the phenomenon, but may reside in intrinsic reasons that no foreseeable progress could gainsay. (Stengers, 1997: 8–9)

In other words, the notion of complexity signals a break in the presumed continuity of different aspects of reality and of the laws that explain them. It signals that no single (or simple) set of questions may be treated as a generalizable norm, in terms of yielding relevant answers for all phenomena. And, last but not least, it applies to situations in relation to which the programme of reductionism, and the cognate aspiration to produce a form of knowledge that is exhaustive and deterministic, does not make sense.

Much more could be said about Stengers's beautiful essay, but I shall limit myself to a number of brief points that will link it to the questions explored earlier in this article. In relation to the theme of vitalism, the notion of complexity is interesting, of course, in so far as complexity is 'usually recognized as almost constitutive of the "living object"' (1997: 13). The theme of complexity allows us to address living beings as original and singular, but not in the sense that their originality stems from an essential difference with respect to the physico-chemical world. Living beings are original and singular in the same way that any complex system is a singularity, on account of its distinctive temporality. Since complex objects, both living and non-living, do represent singularities, it is *both* possible to regard the living as a fundamentally natural object, equal in this sense to the objects of physics and chemistry, *and* impossible to simply extend physico-chemical models – including models of complex objects – to account for the living. For the same reason, living beings may justifiably be regarded as an 'exception' to the principles of physics if the principles we take as our norm are those of classical physics; on the other hand, they will appear as 'normal' singularities if the context of reference is that provided by the notion of complexity itself.

The point I wish to end on, however, concerns a further and more specific link between the notion of complexity and the vitality of vitalism. As we saw earlier, Stengers stresses that what is interesting about complexity is not the particularity of what may be described as complex, but what the notion signifies at the level of conceptual effects. In this sense, the complexification of the world, or the vision of the world as complex, remains relatively uninteresting if it is treated as 'paradigm' that comes in to replace another, without affecting what is understood to be the ethos of scientific knowledge and its relation to the world.

Stengers's emphasis suggests that the theme of complexity can be read as constituting an intellectual demand, an ethical imperative, that is not dissimilar to what Canguilhem addressed through the theme of the 'vitality of vitalism'. Complexity expresses the demand that we acknowledge, and learn to value as the source of qualitatively new questions, the possibility of a form of ignorance that cannot simply be deferred to future knowledge. It is the demand that we acknowledge a sensitivity of the world to our interest in it, and to the forms in which this interest is expressed. This point might serve – in a programmatic way – as the informing principle for an analysis of the use of the term 'complexity' in the context of systems theoretical approaches in contemporary biology. It might offer clues for the purpose of deciphering the vital supplement that the term 'holism' provides, in the context of systems theory, to a notion of complexity that has begun to function as a normative operator, and possibly a source of theoretical self-evidence, in its own right.

Acknowledgement

This article is based on research generously supported by the Alexander Von Humboldt Stiftung. This support is here gratefully acknowledged.

Notes

1. Benton's piece includes a critical review of typologies previously offered by Kemeny, Toulmin and Goodfield, and Jørgensen.
2. See also Wuketits (1985). As Wuketits explains, 'naturalistic vitalism' is a different name for what Timothy Lenoir has called 'vital materialism'.
3. As is well known, it was Gaston Bachelard who first spoke of a plurality of rationalisms corresponding to different regions of scientific practice (*rationalismes régionaux*). In line with this notion, Canguilhem argues that Bachelard's method – and the rigid distinction it implies between positive and negative episodes in the history of science – are particularly suited to the scientific region in reference to which it was developed, namely, the region of mathematical physics and nuclear chemistry (see Canguilhem, 1988: 13). The degree of formalization evident in these sciences is such that the possibility of ambivalence appears relatively insignificant for the purpose of studying their history. It is from within this horizon that 'sanctioned' episodes look like positive contributions to progress, while 'lapsed' episodes look like epistemological debris that constitutes a wholly negative obstacle to the development of knowledge. As Gary Gutting (1989) has rightly observed, Canguilhem thus offers an implicit correction to the notion of 'epistemological obstacle' initially proposed by Bachelard. It is a correction very much in the spirit of Bachelard's own work, that is to say, a specification of the conditions of relevance of Bachelard's historical epistemology. For a fuller discussion of these themes, and for an application to the field of psychosomatic medicine, see Greco (2004a).

4. Here I am using the term 'relevant' in the sense discussed by Isabelle Stengers in the essay 'Complexity: A Fad?' (in Stengers, 1997). I shall come back to a discussion of Stengers later.

5. This proposition is strikingly close to the project developed by Alfred North Whitehead in *Science and the Modern World* (1925). In the reading offered by Isabelle Stengers (2002), Whitehead regarded the life sciences as having been 'handicapped' by their respect for physical explanation, or for 'scientific material-ism'. Whitehead's project in *Science and the Modern World* was to centre the whole concept of the order of nature around the notion of the organism.

6. Lecourt refers in fact to developments in molecular biology, although he cites this text (Canguilhem, 1975: 95–6) to illustrate Canguilhem's 'anthropomorphism' and 'anthropocentrism' (see below).

7. I have developed this critique more fully in an essay called 'The Politics of Inde-terminacy and the Right to Health', see Greco (2004b).

8. On this point, see also Canguilhem's essay 'Machine and Organism' (1992), where machines are (vitalistically) presented as an extension of the organic.

9. See Prigogine and Stengers (1984) for an accessible account of the emergence of a 'science of complexity'.

References

Ackermann, R. (1969) 'Mechanism, Methodology and Biological Theory', *Synthese* 20: 219–29.

Benton, E. (1974) 'Vitalism in Nineteenth-Century Scientific Thought: A Typology and Reassessment', *Studies in the History and Philosophy of Science* 5: 17–48.

Bergson, H. (1911) *Creative Evolution*. Lanham, MD: University Press of America.

Bertalanffy, L. von (1968) *General Systems Theory: Foundations, Development, Applications*. New York: Braziller.

Bronowski, J. (1970) 'New Concepts in the Evolution of Complexity', *Synthese* 21: 228–46.

Canguilhem, G. (1975) *La Connaissance de la vie*. Paris: Vrin.

Canguilhem, G. (1988) *Ideology and Rationality in the History of the Life Sciences*. Cambridge, MA: MIT Press.

Canguilhem, G. (1992) 'Machine and Organism', in J. Crary and S. Kwinter (eds) *Incorporations*. New York: Zone Books.

Canguilhem, G. (1994) *A Vital Rationalist*. New York: Zone Books.

Carlo, W.E. (1966) 'Reductionism and Emergence: Mechanism and Vitalism Revisited', *Proceedings and Addresses of the American Philosophical Association* 40: 94–103.

Cimino, G. (1993) 'Introduction: La Problématique du vitalisme', *Biblioteca de Physis* 5: 7–18.

Dawkins, R. (1988) *The Blind Watchmaker*. London: Penguin.

Emmeche, C., S. Koppe and F. Stjernfelt (1997) 'Explaining Emergence: Towards an Ontology of Levels', *Journal for General Philosophy of Science* 28: 83–119.

Foucault, M. (1989) 'Introduction', in G. Canguilhem *The Normal and the Patho-logical*. New York: Zone Books.

Greco, M. (2004a) 'The Ambivalence of Error: "Scientific Ideology" in the History of the Life Sciences and Psychosomatic Medicine', *Social Science and Medicine* 58(4): 687–96.

Greco, M. (2004b) 'The Politics of Indeterminacy and the Right to Health', *Theory, Culture & Society* 21(5): 1–22.

Gutting, G. (1989) *Michel Foucault's Archaeology of Scientific Reason*. Cambridge: Cambridge University Press.

Haraway, D. (1991) *Simians, Cyborgs, and Women: The Reinvention of Nature*. London: Free Association Books.

Hein, H. (1968a) 'Mechanism and Vitalism as Metatheoretical Commitments', *Philosophical Forum* 1: 185–205.

Hein, H. (1968b) 'Mechanism, Vitalism and Biopoiesis', *Pacific Philosophical Forum* 6: 4–56.

Hein, H. (1969) 'Molecular Biology vs. Organicism: The Enduring Dispute between Mechanism and Vitalism', *Synthese* 20: 238– 53.

Hertogh, C. (1987) 'Life and the Scientific Concept of Life', *Theoretical Medicine* 8: 117–26.

Hoyningen-Huene, P. and F.M. Wuketits (eds) (1989) *Reductionism and Systems Theory in the Life Sciences*. Dordrecht: Kluwer Academic Publishers.

Keller, E. Fox (2000) *The Century of the Gene*. Cambridge, MA: Harvard University Press.

Keller, E. Fox (2001) 'Genes and Developmental Narratives', paper presented at Goldsmiths College, University of London.

Lecourt, D. (1998) 'Georges Canguilhem and the Question of the Individual', *Economy and Society* 27: 217–24.

Lenoir, T. (1982) *The Strategy of Life: Teleology and Mechanics in Nineteenth Century German Biology*. Dordrecht: D. Reidel.

Prigogine, I. and I. Stengers (1984) *Order out of Chaos*. London: Flamingo.

Rapaport, A. (1995) 'The Vitalists' Last Stand', in R.M. Ford (ed.) *Android Epistemology*. Cambridge, MA: MIT Press.

Rose, N. (1998) 'Life, Reason and History: Reading Canguilhem Today', *Economy and Society* 27: 154–70.

Stengers, I. (1997) *Power and Invention*. Minneapolis: University of Minnesota Press.

Stengers, I. (2002) *Penser avec Whitehead: Une libre et sauvage création de concepts*. Paris: Seuil.

Uexküll, T. von (ed.) (1997) *Psychosomatic Medicine*. Munich: Urban & Schwarzenberg.

Weiss, P.A. (1969) 'The Living System: Determinism Stratified', *Studium Generale* 22: 361–400.

Whitehead, A.N. ([1925] 1985) *Science and the Modern World*. London: Free Association Books.

Wuketits, F.M. (1985) *Zustand und Bewußtsein: Leben als biophilosophische Synthese*. Hamburg: Hoffmann und Campe.

Wuketits, F.M. (1989) 'Organisms, Vital Forces, and Machines: Classical Controversies and the Contemporary Discussion "Holism" vs. "Reductionism"', in P. Hoyningen-Huene and F.M. Wuketits (eds) *Reductionism and Systems Theory in the Life Sciences*. Dordrecht: Kluwer Academic Publishers.

Monica Greco is Senior Lecturer in the Department of Sociology at Gold-smiths College, University of London. She is the author of *Illness as a Work of Thought* (Routledge, 1998) and of several articles on psychosomatics and on the ethical and political implications of different forms of medical rationality. She is co-editor with Mariam Fraser of *The Body: A Reader* (Routledge, 2004).

Information and Knowledge

Suhail Malik

IN HIS now venerable 'report on knowledge', Jean-François Lyotard states that technoscientific 'transformations' in cybernetics, communication theory, data storage and transmission, and so on, 'can be expected to have a considerable impact on knowledge'. This has of course become a truism and a reality in the 20 years since the writing of *The Postmodern Condition*, as has the specific determination of this 'impact':

> [Knowledge] can fit into the new channels, and become operational, only if learning is translated into quantities of information. . . . The 'producers' and users of knowledge must now, and will have to, possess the means of translating into these languages whatever they invent or learn. . . . Along with the hegemony of computers comes a certain logic, and therefore a certain set of prescriptions determining which statements are accepted as 'knowledge' statements. ([1979] 1984: 4)

The transformation of knowledge into information demands its codification into a 'certain logic', a certain, determinate, 'operationality'. The movement described here is from knowledge as a mental or 'cultured' human acquisition to its 'exteriorisation . . . with respect to the knower'. Knowledge becomes systemic.

Although Lyotard stresses how this movement leads to knowledge's commodification and mercantalization in capitalism, and to its increasingly central role in ordering power at every level, what it is also highlighted is how this is a movement of knowledge's instrumentalization. Knowledge will itself continue to have meaning only insofar as it will be operational, which is to say in keeping with means–ends and productive logics, categories and functions; that is, insofar as it accords with certain rules, codes and 'prescriptions'. Such a demand limits what knowledge can be; it is a reductive determination of knowledge. And, as is well established by the critical determinations of instrumental rationality in the humanities and social sciences, this reductionism accords with the systemic requirements of efficiency and exchange that do not accord with personal, social or even human needs but are instead directed towards capitalism.

Even disregarding the fatalism of this scenario, the point stands: when it becomes information, knowledge is not free and not in the name of (human) freedom, or anthropic determination, or 'for its own sake'. The move is from knowledge, which has a meaning and value that is intrinsic to it, to information, which has an extrinsically determined meaning and value. In this externalization of what is meaningful, knowledge is determined in terms of exchange-value rather than use-value.

There is a nuance in this argument that needs to be addressed, however. It is not that information requires knowledge to be instrumentalized – as will soon be seen, there may be modes of information that cannot be determined as knowledge. Rather, the point made by Lyotard among others is that the *instrumentalization* of knowledge transforms it into information. The difference to be highlighted here is between the characterization of information as such and, on the other hand, knowledge's instrumentalization as information. What is left begging between these two determinations of how information and knowledge relate to one another is whether information is *per force* an instrumental category. If it is, then the nuance brought to light here collapses and the argument about knowledge has no special place: whatever becomes determined as information becomes instrumentalized and the fate of knowledge is only one such case. However, it seems that there is a stronger connection between knowledge and information than this: specifically, information displaces, if it does not replace, knowledge in the new channels and structures stipulated by technoscientific and contemporary capitalistic determinations. Correlatively, epistemological concerns are displaced, if not replaced, by material-systematic ones. This is, for example, a central contention of Scott Lash's *Critique of Information* (2002) in the description of the collapse of the temporal, spatial or logical conditions for reflection (allied with knowledge) as condition of critique in 'information societies'.

The transmuted identity of knowledge and information can be contested, however. A hesitation about their relation allows for the otherwise intimate knot between these terms to be loosened. With that, some leverage can be had on the question of whether knowledge is in fact, operation, or concept instrumentally displaced by information, or even how the relation and movement from one to the other – if there is such a movement – is to be understood. To be clearer about this, a more precise characterization of information is needed. This is notoriously difficult given the variety of ways the term is understood and operates: as an embodied and material cause (in some genetics), as a signal with a transmissible quotient without determinate location or material/embodied specificity (in cybertheory), as a statistical quantitative property of a system (in communication theory), as an instruction. The divergence in what information means or even is marks it out as an inconsistent category. But it is not incoherent: each of these determinations of information proposes a rationalization of the system in question into (usually linearly organized) channels of control and transmission. In particular, the uncontroversial notion of information as an instruction – with all its implied connotations of programme, design, rule-following, coding, command, and so on – seems to lead directly to its characterization as a type of instrumental rationality, both in its theory and practically. If so, then information is at once an instrumental category and the question of its transformation of knowledge is settled.

Staying at that level of analysis, together with the common theoretical-political positioning and condemnation of instrumental rationality that attends such conclusions, however, foregoes further investigation of what is occasioned within such determinations of system 'operationalities'. That is, certain critical positions are quickly consolidated at this point without further consideration of what something like 'instrumentality' actually amounts to. The following discussion takes up various characterizations of information across a number of disciplines in order to comprehend the effects of such an instrumentalization (if it is one). By examining arguments in theoretical biology, developmental systems, and social theory, the initial stage of the argument establishes that information is a situated event, an event that takes place in a mnemically organized system which, following an argument from developmental biology, is then seen to be epigenetically constituted. This determination provides a basis for a critique of the dominant statistical-quantitative model of information in the physical and mathematically developed sciences, notably as information theory. Dispensing with this model as the primary reference for comprehending information, the argument then turns to various arguments about and in response to 'information societies'. These societies are precisely those which Lyotard addresses in the comments remarked upon above: societies in which information takes a leading role in (de)structuring their economies, knowledges and cultures. As well as Lash's concerns with the reorganization of the spatiotemporal constitution and circulation of knowledge, rationality and (in a general sense) intelligence in such societies, Virilio's argument that information societies are societies of the general accident and Stiegler's investigation of human (social) genesis through what he calls an 'instrumental maieutic' are also examined in view of this general characterization of information. These illustrations accumulate to demonstrate again that meaning is epigenetically constituted – but this time at social and anthropic levels. The last stage of the argument proposes a continuity between this register and the organic epigenesis discussed earlier. This continuity is not that of a staged evolution but the process of information's epigenetic constitution and mutations of (bio)material, symbolic, and social meanings and systems.

It is perhaps worth emphasizing in advance that what is developed and then deployed through these diverse registers is a *general* characterization or theory of information. Such a theory is, on the one hand, required, if the term is not to collapse into the each time specific determinations of the disciplines in which it is deployed; if, that is, the *concept* of information is not to be undone by its disciplinary manifestations. And it is precisely this concept of information and its consequences that this article seeks to establish. On the other hand, working towards this level of conceptual generality and consistency (for the *general* sense of 'situatedness', for example) requires an abstraction from precisely the disciplinary and empirically specific characterizations of information and the constellation of terms that it carries with it. So, though the characterization of information as an event that is situated in a system with an organized memory is developed in the following pages by moving between the biological and social determinations of information, this general characterization does not, however, mean that organic memory and social memory (or personal memory for that matter) are the same thing or are unproblematically identifiable, or that what information is, in fact, is the same thing in every instance. In fact, the argument here suggests that it cannot be since what and how the mnemic organization is for each developmental system is, precisely, different and particular. The concern here is not then to specify what, for example, the mnemic organization is, and so what the specificity or actuality of information is, in every case. For all that, such specificity is observed in the later section on information societies since what is being constructed here through the examples considered is an understanding of the centrality of information, not just in the formation of certain kinds of society, but, more intrinsically, as a condition for the constitution of complex societies at all.

What appears through these arguments is that rather than being the reduction of knowledge, information involves a complexification of meaning and systems, a complexification that can now be attributed to the operation of information with some specificity. In particular, there is a reversal of the conventional critical attributions noted above: it is not that information is the latest or only a particular mode of instrumentalization (of knowledge, say) but rather that instrumentality is a particular stratum of informatic operationality, one that determines and is determined in an anthropotechnical complex (and which therefore precedes consciousness as the condition for knowledge in its anthropically derived sense). The received critique of instrumentality is thus seen to be a disavowal of this complex mnemic organization that is central to the development of all orders of living systems and societies; central, that is, to production from pre-cellular-organic to post-industrial levels.

The Situated Event

In her critique of the reductionist concept of information in biology, propounded for example by neo-Darwinianism,[1] the developmental systems theorist Susan Oyama proposes that the information in a living system is not reducible to the gene – or any other one element – as sole site or causal control of heredity. Rather, what counts as information depends upon the conditions of the system under question at any particular time:

> The 'informational' function of any influence is determined by the role it plays in the developmental system as whole. Regularity of gene function is thus a

result of developmental regularity as well as a cause of it. . . . What is crucial is not *permanence*, but *availability at the appropriate time*. Persistence is beside the point in accounting for reliability. (Oyama, 2000: 84)

The generational regularity and variation of the system – the information conveyed in the system and by it as regards its transmission – then depend on the system as a whole at the time at which the transmission is taking place. Oyama's extended and *systemic* notion of information develops Gregory Bateson's deceptively simple definition of information as 'a difference that makes a difference' (Bateson, [1972] 2000; Oyama, 2000: 67). Despite the abstraction of this definition, it immediately proposes a relational determination of information. As Oyama puts it:

This invites questions: a difference in what (What are you paying attention to?), about what (What matters?), for whom (Who is asking, who is affected?). Asking these questions leads us to focus on the knower, a knower who always has a particular history, social location and point of view. (2000: 147)

The point here is that Bateson's definition of information as a difference that makes a difference, and Oyama's extension of it as being correlated to the 'role it plays in the developmental system as a whole', propose a set of relations and a history as the conditions for information. To say that there is a difference is to posit a relation and a history, a memory (of whatever sort, technical or natural), with respect to which a difference can be determined. But for that difference to be information rather than just a difference, it must also *make* a difference – it *alters* extant relations and memory. As information, the difference in question presumes *and* generates a specific organization of relations and memory. By contrast, as will be seen below, the statistical-quantitative definition of information is that it is the reduction of uncertainty. This corresponds to the specification of fewer organizations of order in the range available at the moment that the information takes place. Even at this early stage of the argument, then, it can be noted that such a definition is a limitation of the wider – qualitative, ontological – definition proposed by Bateson and others following him.

By virtue of a difference recognized as such, a relation is constituted; in making a difference, information proposes a system and organization in alteration. This latter point is re-articulated by a systems theorist with different concerns: Niklas Luhmann:

By information we mean an event that selects system states. This is only possible for structures that delimit and presort possibilities. Information presupposes structure, yet is not itself a structure, but rather an event that actualizes the use of structures. ([1984] 1995: 67)

As a relational term, information presupposes organization, requires systemization: 'Information is always information for a system' (Luhmann, [1984] 1995). The presupposition of structure for there to be information is what leads to characterizing it as necessarily 'operational' and even instrumental. However, this very determination of information as intrinsically systemic, as 'operational', cannot be separated in principle from what is often held to be contrary to such reification in the critical parlance: its eventhood.

Information is an event for three main interconnected reasons. First, it makes a difference – it is event-like in that it alters the state of the system. Second, it cannot be repeated – the repetition of information (an event) is not further information but no information (not an event as such). It is event-like in this respect – or eventive[2] – in that each instance of information happens only once. Third, it temporalizes and historicizes the system – its alteration produces a 'before' and an 'after' for the system, either for the system itself (as an experience) or for an observer (as a history). Its evention here is that in a certain way it happens in time – if, that is, it does not generate the experience of the actuality of time. As Luhmann puts it, and as the example makes clear:

> a piece of information that is repeated is no longer information. It retains its meaning in the repetition but loses its value as information. One reads in the paper that the deutschemark has risen in value. If one reads this a second time in another paper, this activity no longer has value as information (it no longer changes the state of one's own system), although structurally it presents the same selection. The information is not lost, although it disappears as an event. It has changed the state of the system and has thereby left behind a structural effect; the system then reacts to and with these changed structures. (Luhmann, [1984] 1995)

Through its repetition, information 'retains its meaning' but 'disappears as an event'. A piece of information alters the state of the system, changes the organization in its actuality, and that alteration is retained as a (new) meaning, which is to say an altered mnemic organization of terms and their relations. Equally, the repetition of the information does not then further transform the system: since the repetition does not alter the mnemic organization of the system, it disappears as information (unless the information is 'forgotten', i.e., the system returns to an earlier state). Conversely, if the system or its internal meanings are not altered by what is ostensibly a piece of information, it is not in fact information. Its meaning/effect is already within the system and is consolidated. Indeed, its meaning – and the meaning of the system to itself – remain intact *because* it is not information.

Information is thus a *situated event*, an event that *generates* meaning in a system or for an organization. As a generation of meaning, this characterization goes beyond Luhmann's limitation of information to a 'selection' of system states. Information events meaning. In this movement of retention and alteration, the system has an open history. This is why the systems under consideration can rightly be called 'developmental systems', for which information is then a central concern (Oyama, 2000). And it is also why, in a context to which we return later, even if capitalism is taken to operate systemically, monologically, or axiomatically it is nonetheless (internally) eventive in its informatic extension (information being understood here beyond the narrow determination of it through information [i.e., electro-computational] technologies).

There are thus two ways in which meaning occurs in a system. One is as information, which is also the alteration of the system, the active or passive shifting of the specific conditions of organization, its transduction and evention. The second is as a meaning that is not in the temporalization of the event of information but in the established memory, structure or organization of the system. It is a meaning that is intrinsic to the system and intrinsic to itself, maintained and preserved in its significance in the system. What needs to be emphasized, however, is that without this latter dimension of meaning, there can be no information as such since it is only this more or less established memory that provides the 'situation' for the situated event that is information. This mnemic conditionality and fate of the operation of information are not just limited to social, cultural or technical registers. Such a condition also plays a key role in organic developmental structures, as Marcello Barbieri argues in his inventive challenge to the established models of both genetic determinism and also developmental theory in embryology.[3] Barbieri's principal concern is to explain how a cell increases its own complexity in its development, and how that increasing complexity converges, 'in the sense that the outcome is neither random nor unexpected' (2003: 3); that is, given equal developmental conditions, cells of a certain type develop pretty much in the same way each time. The problem confronted by Barbieri, as it is by gene determinism and embryological developmental theory, is how such a convergent increase in complexity can take place since the information required to organize the development of a cell, the phenotype, is not contained in the cell's genome, as the doctrines of molecular biology have asserted for the past century or so:

> the information of a gene is determined by the order of its nucleotides, pretty much as the information of a word is due to the order of its letters. In both cases information corresponds to the order of elementary units along a line. Genetic information is therefore a *linear* quantity, but the function of proteins is determined by the arrangement of their amino acids in space, i.e., by their *three-dimensional* information. Clearly genes are not transporting all the information that is going to appear in proteins. Where then does the missing information come from? (2003: 30)

In other words, the genomic model of development is insufficient to explain or determine cellular development. Barbieri proposes that the increasing complexity of the living system is attributable to an *epigenetic* reconstruction, a reliable development of internally meaningful and complex structures that exceeds the genetically determined information within the system since new properties develop in the system at each stage of its development:

> The information difference that exists between the linear order of polypeptides and the three-dimensional order of proteins can be illustrated with a

simple example. The linear order of 100 punctiform amino acids is specified by 100 coordinates, while their three-dimensional organisation requires 300 coordinates (three for each amino acid). Protein folding, or self-assembly, amounts therefore to adding the 200 missing coordinates to the 100 coordinates provided by the genes. And since the complexity of a system is determined by the number of parameters that are required to describe it, it is clear that protein folding is a phenomenon that produces *an increase of complexity*.

In embryonic development . . . the term *epigenesis* has been used to describe the increase of complexity that takes place in a growing embryo, but that term can be generalized to any other convergent increase in complexity, and we can therefore say that protein folding is an example of *molecular epigenesis*. (Barbieri, 2003)

In this sense, epigenesis is 'a process [of] reconstructing a structure from *incomplete information*' with a convergent increase in complexity of that structure (2003: 71). The relevance of Barbieri's work to the present argument is that such epigenesis is in fact arranged according to 'organic codes' that (following a model borrowed from linguistics) determine a meaning for the living system at the level of its biological function and development:

A code can be defined as a set of rules that establish a correspondence between two independent worlds. The Morse code, for example, connects certain combinations of dots and dashes with the letters of the alphabet. . . . The extraordinary thing about codes is that a new *physical quantity* appears in them, since they require not only energy and information but also *meaning*. . . . The words of language may seem arbitrary if taken one by one, but together they form an integrated system and are therefore linked by community rules. Codes and meanings, in other words, are subject to collective, not individual, constraints. Codes have, in brief, three fundamental characteristics:

(1) They are rules of correspondence between two independent worlds.
(2) *They give meanings to informational structures.*
(3) They are collective rules which do not depend on the individual features of their structures. (2003: 94; emphasis added)

Organic codes are then the internal *meaning* of the living system (which does not even have to have reached the cellular stage of development). These organic codes/memories 'situate' or, in Barbieri's terms, contextualize (2003: 111), the function of any molecule in the cell (including the genome), enabling it to inform the development and (internal or external) function of the living system. The codes give the elements of the living system an organic meaning. Barbieri's contention is that these codes belong to neither the genotype nor the phenotype of the system but are in fact their common logical and historical condition,[4] and even artifactualize them (2003: 160).

This theory of the development of living systems as being conditioned in the first instance by organic *meaning* – whence Barbieri's 'semantic biology' – surpasses the limitations inherent in both molecular biology and embryology in accounting for the development of living systems, whether they have genomic components or not. The key aspects of concern here are the following: first, information constitutes an *epigenetic* (or, in Oyama's terms, non-persistent) memory for the system. The organic codes of the cell are therefore dual:

> A memory is a deposit of information, and we can give the name organic memory to any set of organic structures that is capable of storing information in a permanent (or at least in a long-lasting) way. The genome, for example, is not only a hereditary system but also an organic memory, because its instructions are not only transmitted to its offspring, but are also used by the organism itself throughout its life. We can rightly say, therefore, that the genome is the *genetic memory* of a cell.
>
> The state of determination [for the 'histological fate' of a cell, its differentiation] has also the characteristics of an organic memory, because it has the permanent effects on cell behaviour, but it has an *epigenetic* memory, i.e. a memory which is built in stages during embryonic development by epigenetic processes. We conclude therefore that embryonic cells have two distinct organic memories: the genetic memory of the genome and the epigenetic cell memory of the determination. (2003: 114)

That is, there are (at least) two mnemic registers that determine the developmental order and regularity of a cell, not just the one stipulated by the genoinformatic determinism of neo-Darwinian molecular biology. Second, then, the dual organic memory of the cell gives the linear information contained in the genome its meaning in the developmental process. And the more general inference is that the organic memory of the cell gives whatever information is contained in the cell – either in its linear coding or in its spatiotemporal ordering – functional effectivity:

> Without [an epigenetic] memory it would not be possible to obtain a convergent increase in complexity, and the real logic of embryonic development is precisely that kind of increase. . . . Cell memory is a key structure of embryonic development because it is essential to the convergent increase in complexity that is typical of development. (2003: 120)

Without such a memory, whatever information there is in the developmental system would be insufficient to determine the very development it is meant to explain, for example, in the doctrine of modern genetics.

In this mnemically organized production of information and the informational production of memory, the cell – the primary model of a living system – is therefore a mnemic *and* temporalized system. Though Barbieri makes a distinction between 'informatic processes, where only energy and information are involved' (2003: 95–6) – which are the systems considered by the physical sciences – and 'semantic processes, where rules appear which add meaning to information' (2003: 95), it is the case that the information in a system events nothing, has no effects, without the latter.

Barbieri's argument from – and to – biology demonstrates the principal argument here at the level of nature and its sciences: information requires meaning and memory (here, the organic codes) in order for it to be information *and*, given that requirement, information generates further meaning. What is proposed is a more comprehensive account of the conditions for both models that recognizes the condition of an organization of (here, organic) memory – and therefore (organic) meaning – for there to actually be information at all. Information is meaningful in the system. Or, meaning is dual: it is mnemic and informatic.

Critique of the Statistical-Quantitative Determination of Information

Having established the two dimensions of how meaning occurs in a system or organization, we can come back to the question of the relation between knowledge and information. It can now be seen that the instrumentalization of knowledge as information, Lyotard's concern in *The Postmodern Condition*, is the putting into time and alteration, the 'eventing', of the memory, organization and meaning of a system (even to itself: information can come from within a system as well as from without). The challenge of this evention to the received account of meaning is that in the proto-Platonic account the principal (anthropic) securing and ordering of meaning is *knowledge*. Comprehended as the established (if not, as Plato writes in the *Meno*, sempiternal) memory and organization of a system, knowledge is the conservation of (the meaning in) a system (including its own meaning for itself) and information is the alteration of that system and, correlatively, in however limited or extensive a way, its meanings.

The distinction between the principal terms in question here is then clear: knowledge is mnemically and organizationally confirmed meaning while information is the temporalization of meaning, the generation and emergence of meaning and system, of organization. And it is this proto-Platonic stabilization of meaning as primarily knowledge (with its consequent stabilization of information in reference to meaning) that lies at the heart of the communication theory account of information in the physical and mathematically derived sciences, as well as in the discourses that claim their authority. This is unintentionally demonstrated in Fred Dretske's *Knowledge and the Flow of Information*. Dretske identifies the alteration of a situation by information – its evention – as 'learning' (1981: 45 and passim), and gives information the following 'nuclear' sense: 'a state of affairs contains information about X to just that extent to which a suitably placed observer could learn something about X by consulting it' (1981: 45). Suffice to say that Dretske's 'state of affairs' and 'suitably placed observer' all testify to the 'situatedness' of information that is being emphasized here. Learning, then, would be another name for the temporalization of meaning that is evented by information. However, Dretske stabilizes the condition of meaning, and consolidates the communication theory account of information with it, by arguing that this 'ordinary, semantically relevant, sense of information . . . is something to be distinguished from the concept of *meaning*' (1981: 46), suggesting instead that the relevant criterion for information is 'whether it can provide an illuminating account of that commodity capable of yielding *knowledge*' (1981: 46; emphasis added). Dretske's argument is that knowledge is an absolute category since it is not a matter of degree (in our terms, it is consolidated):

> factual knowledge . . . does not admit of . . . comparisons. If we both know that the ball is red, it makes no sense to say that you now this better than I. . . . If a person already knows that the ball is red, there is nothing he can acquire that will make him know it better. . . . In this respect factual knowledge is *absolute*. (1981: 107–8)

For Dretske, the absolute character of knowledge stems from the absolutism of information 'on which [knowledge] depends', on information's eventual redundancy. Information can be redundant because it contributes nothing to knowledge:

> Information itself is not an absolute concept . . . since we can get more or less information *about* a source. Information *about s* comes in degrees. But the information *that s* is [the knowledge] *F* does not come in degrees. . . . Once the information that *s* is *F* has been received, there is *no more* information to be had about whether or not *s* is *F*. Everything else is either redundant or irrelevant. (1981: 108–9)

Dretske is in his own way articulating the point we saw Luhmann make earlier: the repetition of the same piece of information is not information. The difference from Luhmann's characterization, however, is that for Dretske this redundancy arises not because information can happen only once (in its evention) but because knowledge as knowledge is immutable and because further information can add nothing to knowledge. The Platonic assumption here is clear. Meaning for Dretske is to be filled and completed rather than mutated. What follows (in a perfectly Hegelian way) is that information is determined by and directed towards such a stabilized or fulfillable knowledge – as Dretske puts it, 'no more' information can modify knowledge once there is knowledge. Hence their common absolute character.

Meaning *qua* information can thus be exhausted in knowledge – and therefore quantified: the certainty in knowledge offers a 0 (no knowledge) and a 1 (knowledge) by which the relevance and transmission of information can be calibrated. Dretske establishes the mathematization of information by communications theory on this basis. What is measured between the 0 and 1 is not the information itself or the knowledge it generates but the material and practical context in which that knowledge and information take place:

> To know, or to have received information, is to have eliminated *all relevant alternative possibilities*. These concepts are absolute. What is not absolute is the way we apply them to concrete situations – the way we determine what will qualify as a relevant alternative. . . . Knowledge exhibits [a] pragmatic character *because* a communication system, *any* communication system, presupposes a distinction between a source *about which* information is received and a channel over which this information is received. The source is the generator of (new) information. The channel is that set of existing conditions that generate no (new) information. . . . It is in the determination of what constitutes a relevant alternative, a determination that is essential to the analysis of *any* information-processing system (whether it results in knowledge or not) that we find the source of that otherwise puzzling 'flexibility' in our absolute cognitive concepts. (1981: 133–4)

Having evacuated meaning from information by premising both on the absolutism of knowledge, what is being measured and operationalized in communications theory are the 'relevant alternatives', the possibilities, as is well attested, that are proximate to a piece of information or knowledge (here, a signal) in a 'concrete situation'. In other words, what communications theory is concerned with for Dretske is not information as such, nor knowledge as such, but the pragmatic conditions for the transmission and 'relevance' of information. The communications theory of information is then a pragmatics.

But this solution runs directly into the problem which it is supposed to solve: the 'relevant alternatives' in a 'concrete situation' speak *in fact* to the situation of the information that is being transmitted. To that extent, it is to speak about the range of actual meanings of information rather than what Dretske would call a knowledge of it. What other gauge can there be for the '*relevant* alternatives'? What is therefore being accounted for in such determinations of information is the paradoxical formulation of a range of meanings whose meaning has been evacuated. Though this abstraction is the condition for the mathematization of information (and in fact all it is concerned with), it is also the reason why what is addressed through such formulations is not in fact information but an absolute knowledge of situations.

It is no great surprise to see that the mathematical abstraction of information is also its (proto-Platonic) epistemological securing in its being premised on a consolidation in and of knowledge taken to be absolute rather than informatically transmuted. As is firmly established from the Shannon–Weaver model of information and its subsequent development, into cybernetics and A-Life, for example, it is a probability analysis indicating that a more or less situated transition can take place in the alteration from one state to another of a system. But this rendition belies the very conditions of information in that it negates the situatedness, the mnemic organization, in which information has its always unique operation and temporalizing evention.

In every sense, then, actively and by default, the mathematical epistemology of information refutes the meaning of information. This does not present any difficulty to that model of information because it has decreed from the off and has striven to establish that information has no meaning *per se*. The only question that remains is what then it is dealing with if not the very absolutely or sedimentation of knowledge that information in fact alters. Likewise, organic memory as Barbieri speaks of it – as the primary condition of meaningful information – is abstracted away by the doctrine of genetic determinism, predicated as the latter is on the notion of asemantic information as the sole condition and control for development. What is left for consideration in both cases are the statistical and quantitative properties of signal transmission rather than the semantics of the signal. And what is evacuated from such considerations is that information is a situated occurrence – situated in and by a mnemic organization. This limitation means that the consideration of living systems by the physical sciences or on their model – such as in terms of purely 'informatic processes' – is in fact incapable of comprehending them at all.

Information – Society

If, by contrast, information dominates knowledge as the primary condition of meaning, then epistemological securing is displaced, if it is not replaced, by (an ontology of) systemic evention. Three analyses of this displacement have recently been put forward, each of which addresses the characteristics and impact of the 'information age', meaning primarily the era of the proliferation of electro-computational networks. Their brief sketch-review will be instructive in bringing the characteristics of the informatic dimension and operation of meaning more clearly into view.

First, Lash's 'critique of information' highlights the duality of meaning in focus here. Lash argues that there are two types of information in the information society. The first is the production of discursive or analytic knowledge which replaces labour production (2002: 141–4). The second type of information 'has to do with the unintended consequences of the first type of information' (2002: 144), namely information overload and disinformation, 'the out of control anarchy of information diffusion' (2002: 146) that makes reflection impossible and leads to an irrationalization. This 'risk society' thesis leads to another central contention of Lash's book – that 'the critique of information is information itself' (2002: 220 and passim). Lash's twofold account of information society is consistent with the duality of meaning as knowledge and information that is being highlighted here. There is nonetheless an important difference in terminology that needs elaboration since it is also a difference in orientation. What Lash calls *dis*information is, in our terms, information in its eventing, information *as such*. And what he calls information we would say is knowledge, as is partly acknowledged in Lash's description of knowledge-production. Though Lash is explicit in insisting that the 'essence of the information' is 'the contradictory pair, the undecidable of *information-and-disinformation*', such that 'disinformation converts just as readily back into information as the reverse'

(2002: 154) and there can be no separation between information and disinformation, it is nonetheless important to emphasize that with the analysis proposed here the 'risk' effects of the information society are not 'consequences' of information but are the operation of information itself. Risk, the meaning of risk, is occasioned with information because information involves the transmutation of existing conditions. The alteration to knowledge that follows is just the informatic dimension of *meaning*, even if it is not meaning*ful* and even countermands meaningfulness (as knowledge). In this way, the 'anarchy' of information is institutionalized – mnemically organized – to some degree or other. It is then not so much that 'the critique of information is information itself' but that information is the critique of *meaning*, the vector of meaning's transmutations. The move from founding and orienting social order and meaning on the basis of secure knowledge, whatever local modality that securing may take, to 'information societies' (the move of capitalism, arguably) is a move that promotes a society turned towards evention and the alteration of meanings. But it does not spell the collapse of meaning – only that meaning is not established or, pushed to its limit, establishable. It is contingent – on new information.

Second, Paul Virilio's 'accident thesis' also articulates the risk of information at the sociotechnical-global level. Virilio adopts Aristotle's categorical distinction between substance, which is 'absolute and necessary', and accident, which is 'relative and contingent', to inscribe a logic of global history: 'we can now equate "substance" with the beginning of knowledge, and the "accident" with *the end* of that philosophical intuition initiated by Aristotle and a few others' (2003: 25). That is, the accident spells the end of knowledge. And, for Virilio, this is a global concern because the sociotechnical developments of modernity have industrialized and eventually globalized the production of the accident such that it is, in another version of the risk society thesis, the habitus of the contemporary world:

> if *speed* is responsible for the exponential development of the *man-made accidents* of the twentieth century, it is equally responsible for the greater incidence of *ecological accidents* (in the various cases of environmental pollution), as it is for the *eschatological tragedies* that loom with the recent discoveries relating to the computing of the *genome* and biotechnology.
>
> Whereas, in the past, the *local accident* was still situated (*in situ*) – the North Atlantic for the *Titanic*, for example – the *global accident* no longer is, and its fall-out extends to entire continents. Waiting in the wings is the integral accident, which may, some day soon, become our only habitat. (2003: 24–5)

The accident, as Virilio describes it, is of the order of the event, and it undermines the securing of a social or personal order on the basis of knowledge. It can be equated with the evention of information as it is being elaborated here. What is at stake for Virilio in this re-orientation of meaning at all levels is, however, the 'catastrophic' abolition of knowledge, including the mode of knowing that is consciousness. Virilio issues the dire warning:

the loss of consciousness of the accident, and of the major disaster, would amount not just to thoughtlessness, but to madness – the madness of voluntary blindness to the fatal consequences of our actions and inventions (I am thinking in particular of genetic engineering and the biotechnologies). . . . We would see the fatal emergence of the *accident of knowledge*, of which information technology may well be the sign by the very nature of its undoubted advances, but also by the incommensurable damage it has done. (2003: 7)

The promotion of the accident by information technologies is reiterated by Virilio's observation that 'it is the intensive use of powerful computers that has facilitated the decoding of the map of the human genome, thus fostering the fateful emergence of the genetic accident' (2003: 27, n.2).

The 'accident of knowledge' Virilio is so concerned with is the de-substantialization of knowledge. But this is only the effect and operation of what is here being called the evention of information, the vector of meaning's mutation. Comprehending Virilio's accident thesis on this basis, what is apparent is that Virilio's catastrophic-eschatological conclusions and apprehensions are derived from an anxiety about the de-securing of knowledge as the condition of meaning by information. That is, Virilio implicitly desires knowledge as a secure basis of meaning and thus relies upon a certain Platonic sensibility for the logic of his argument and anxiety. Without such a premise, the increasing prominence of information as the continued production of meaning, the 'emergence of the accident of knowledge' as Virilio puts it, is nothing absolutely new, nor does is necessarily result in the 'fatal' habitus of the global accident. On the contrary, what can be affirmed is that insofar as meaning is informed, it is produced by accident, so to speak. The accident of information annihilates the world of meaning if and when its evention overwhelms the mnemic organization or (natural, social or technical) developmental system in which it occurs. If, however, the accident cannot be registered in any mnemic organization or system, it cannot be comprehended as information. In this case, the accident has no meaning. It is an event without evention.[5] Such an event-accident cannot be ruled out, mainly because it is precisely that which is beyond any power or rule. However, the identification of such an accident with information depends upon the prior repudiation of information as constitutive of meaning – which is to say that, in a kind of contemporary Platonism, it presumes knowledge as the sole condition for meaning. What would then have to be explained but could not be is the history and development of meaning without its accidents, without its epigenesis.

Third, in *Technics and Time* Bernard Stiegler ([1994] 1998) proposes an account of the epigenetic emergence of the anthropo-technical complex: the coming into historical and temporal being and meaning of the human – which is to say a relation to the past and the future – *through* its technical 'accident'. Stiegler's manifold and complex argument cannot be reproduced here in all its dimensions. Only one aspect of its argument from (a very specific) palaeo-anthropology will be summarized here. Stiegler takes up the work of Leroi-Gourhan on the early emergence of the human from its animal ancestors, a transition that Leroi-Gourhan argues is one of a change in skeletal frame to the upright posture. This enables the enlargement of the skull and the flattening of the face which in turn allows for the increase of brain size. All this is relatively uncontroversial. However, Leroi-Gourhan's importance for Stiegler (and Derrida before him) is his thesis that this process of hominization or brain-enlargement and reorganization is *conditioned* by the technical instrument, notably the accompanying evolution of the flintstone:

> From the Zinjanthropian to Neanderthal man, a cortical differentiation as well as a lithic differentiation is effected, extending from the flaked pebble and the laurel leaves of the Neanderthalians to the [flint] biface [of the early axe]. We submit that . . . cortex and equipment are differentiated *together, in one and the same movement*. The issue is that of a singular process of structural coupling in *exteriorization* that we are calling an instrumental maieutics, a 'mirror proto-stage' in the course of which the differentiation of the cortex is determined by the tool as much as that of the tool by the cortex; a mirror effect whereby one, looking at itself in the other, is both deformed and formed in the same process. (Stiegler, [1994] 1998: 157–8)

The importance of this argument is that it undoes the traditional mono-determination of the technics as invented by the human, in terms of either the technical instrument emanating from the human (mind or body) or of serving human ends, that is, technics designed by and for the human. Rather, the evolutionary and social development of the human is co-conditioned by technics. With the human, then, the process of the development of life is 'exteriorized' from its biological determinants and evolution into technical artefacts that at once continue the development of the human and, importantly, because it is taking place technically and thereby effectuates a mnemic trace that is external to any one individual, that development is *socialized*. The continued development of the human (a form of life) is then not (just or predominantly) biological; it is primarily social and technical. Technics – and the society it inaugurates – are the specificity of the human ([1994] 1998: 157). At the same time, it is 'the pursuit of the living by other means than life' ([1994] 1998: 135 and passim).

The key point for Stiegler throughout is that the technical instrument is a memory of accumulated knowledge that is *external* to the biologically determined aspects of human development. Two corollary points are relevant here: first, human development – understood from now in the sense of the anthropotechnical complex – happens through the development of these external memory supports, of technics and its societies. It can therefore take place through those external supports at rates faster than those allowed for by biological development. Second, such a development is in every case an accumulation of knowledge, memory and learning from an earlier generation rather than an immutable code. For any generation the tools of its predecessors are 'already there' as a technical maieutic. In other words, anthropotechnical development is an epigenetic process. Moreover, since this process speaks to the specificity of the human in its distinctness to other forms of life, it is in fact an epiphylogenetic process:

> The point here is to focus on the originality of the epigenetic process that is put in place from the moment of the appearance of tools, insofar as they are conserved in their form beyond the individuals producing or using them.
>   . . . Epiphylogenesis, a recapitulating, dynamic, and morphogenetic (*phylogenetic*) accumulation of individual experience (*epi*), designates the appearance of a new relation between the organism and its environment, which is also a new state of matter. If the individual is organic organized matter, then its relation to the environment (to matter in general, organic or inorganic), when it a question of the *who*, is mediated by the organized but inorganic matter of the *organon*, the tool with its instructive role (its role *qua* instrument), the *what*. It is in this sense that the *what* invents the *who* just as much as it is invented by it. (Stiegler, [1994] 1998: 176–7)

The pertinence of Stiegler's argument to the present discussion is in this notion of epiphylogenesis, of the who-what – or anthropotechnical – complex of invention. In this complex, the mnemic organization of meaning is technical as well as organic. The human in its biology – in particular, through the plasticity of its brain as regards species determination of behaviour (Stiegler, [1994] 1998) – is a mnemic organization by which technics takes on its meanings in the instrumental maieutic. Equally, technics is also a mnemic organization by which human action takes on meaning (gestures, language). As Stiegler puts it in speaking about modalities of memory: ' what takes place here . . . is the passage from . . . the genetic to the nongenetic' ([1994] 1998: 138). Which is also to say a passage in mnemic conventions, or *a passage in meaning*. In other words, what Stiegler presents in the hominization of life, which is its technicization, is the emergence of a mutation in meaning from the one determined by biology: the system of meaning that constitutes life is altered by the instrumental maieutic of technics through the human. Hominization, the development of the human, is in this sense a process of information. The human, more exactly, the human brain and its society, is the site for information. Information is not then external to the interests of the human (as Virilio's doom-laden scenario pictures it, for example). It is rather another name for its continued development, and the development of (its) meaning, as a complexly constituted phylum.

But here it is not just information in the limited sense of the alteration of a certain modality of mnemic organization. What Stiegler proposes is rather the constitution of a mode of mnemic organization altogether other than the biological one, hence its 'rupture'. This goes beyond the description of information as the alteration or eventing of a mnemic organization. Stiegler's key term 'invention' thus seems a more appropriate term than evention to capture the break-passage – the schiz – in modalities of memory that Stiegler affirms in the process of hominization. But we can note that the distinction Stiegler draws between human development through the sociotechnical exteriorization of memory and the rest of organic life – the specificity of the human – is somewhat weakened by the 'organic codes' Barbieri speaks about. Because what Barbieri proposes is that even at the level of biological formation of the cell, a process of epigenetic development is underway that includes the genetic code as but one of its determinants. Hence, any distinction between genetic determination and epigenetic formation is undermined at the base biological level of the cell if not below it. The distinction between the human and the rest of life made by Stiegler on the basis of the former's epigenetic development and the latter's more completely genetic formation cannot then be sustained. The continuity rather than rupture between anthropotechnical life and life in general is further confirmed by Barbieri's argument that even at pre-cellular stage of the production of ribonuclear proteins, the presence of coded polymers requires the production of polymer units *external* to the organic codes that generate them. That is, there is a production of molecules 'from without' (2003: 160) – an artificial production. The organic codes act in this sense as a technical memory at the level of pre-cellular organic production.

Instrumentality Reviewed

Stiegler's 'instrumental maieutic' as an epigenesis vectored through an external mnemic organization is not then restricted to the emergence of the human as anthropotechnical complex. Rather, 'it structures all levels of the living' as he puts it in a commentary on Derrida's notion of the *gramme* ([1994] 1998: 137). If, then, epigenesis can be identified with the process of information's evention of meaning, it can be said that information – the alteration of meaning – is at the heart of the development of life from organic to inorganic organized matter, from base levels of the biotic to the technical, for which the human is then one site *among others*. What is in transition, what is in continuity, through this rupture is meaning, from organic meaning to socio-technical – i.e., symbolic, cultural and operational – meaning.

Some caution, however, needs to be observed about the continuity of the epigenetic constitution of meaning from pre-cellular-organic to social informatic processes. As stated in the introductory comments above, the interest here has been in developing a *concept* of information, and this entails a certain necessary degree of abstraction and therefore of generality. The situatedness of information, however, means that *what* information is, and what meaning/mnemic organization is at any level (or even sublevel), in any situation, is distinct and demands differentiation from what it is at any other. That is, although it is argued that the process of informatic-epigenetic constitution described here runs from pre-cellular-organic to post-industrial-social conditions, this ought not to be confounded with a developmental continuity between them. It is an argument for a conceptual consistency in how 'development' (which is to say, changes in meaning and systems) happens: informatically. This conceptual continuity is, to repeat, instanced in each and every case differently because of information's (and therefore meaning's contingent) situatedness.

The 'instrumental' relation in the limited sense that is Lyotard's concern (among many others) is this instrumental maieutic understood from an anthropocentric viewpoint. The received critique of instrumentality – instanced above by Virilio in particular – repudiates the general operation of information as constitutive and, at once, (de)structuring of human meaning if not, differently, life and, differently again, social order. Understood otherwise, however, 'instrumentality' in the limited sense is but the hominoid – which is to say, the *inextricably* anthropotechnical – stage of the development of life in its meaning, what Stiegler calls its 'maieutic' and what is here being called its evention. It is information that alters, and continues to alter, the mnemic organization of living and social – henceforth unstable – systems. That is, instrumentality is *intrinsic* to the constitution of organic life, anthroponoetic operation, and social (dis)structuring. It informs life in general.

At any level, then, information as the eventing of meaning cannot be separated from the operationality and instrumentalization Lyotard speaks of. In fact, it is the one process: information is operational in a way knowledge is not to the extent that it alters the system and selects states that are open and mutable in its structure. This is not to say that the informatization of knowledge in the mode of computerization does not mean the stipulation of 'a certain logic' and a 'certain set of prescriptions', as Lyotard put it. Only that this mode of information is but *one* technical determination of information and not the only one. And that, perhaps even in this case, such an informatization is no less a way in which the mental and social mnemic and organizational systems of the human in modernity are reconstituted not on the basis of its maintenance, its memory and sustained organization, that is, on the basis of what is *known* of it – which is what a logic and prescriptions more traditionally codify – but in terms of generalized evention.

Notes

1. The reader unfamiliar with biological terminology may require the following in order to make their way through this section. Standard theories of heredity in modern biology split the organism into two distinctive aspects: the phenotype and the genotype. The *phenotype* is the form of the organism; loosely speaking, its 'body' and physical characteristics. The organism's *genotype* is its hereditary material, which is localized in most contemporary biology and medicine to the gene, which is in turn found in the cell's chromosomes. This distinction between genotype and phenotype was made initially in 1909 by Wilhelm Johannsen (who also coined the term 'gene') in the attempt to account for the Mendelian model of inheritance in distinction to the Darwinian generality of natural selection (which includes the phenotype in its account of species development). After much initial dispute, this distinction was consolidated in biology in the 1940s (see Barbieri, 2003: Chapter 8; Depew and Weber, 1995: Chapter 9).

The most important point to be made about this distinction for the purposes of the present discussion is that the dominant doctrines of modern biology, neo-Darwinism and its experimental practice, molecular biology, locate all (or nearly all) of the developmental information of the organism in the genome, i.e., in the gene which is therefore a kind of 'map' of the organism as a whole. It is there proposed that control and modification of the gene will *on its own and all other things being constant* result in a specifically modified organism (the phenotype). The genotype is supposed to 'suppl[y] the fundamental pattern of the organism' (Oyama, 2000: 16); it is the 'biological software' of the organism (Barbieri, 2003: 25), 'a deposit of instructions and therefore . . . potentially capable of carrying the project of embryonic development' (Barbieri, 2003). What is important in this determination of organic development as a background to the present argument is that the phenotype is degraded if not ignored as a factor in development, never mind the environment in which the organism exists. That is, the 'situation' of the genome is ignored and the information controlling development (as this doctrine would have it) is attributed exclusively to a material and substantial location – as a design – in the gene – whence the 'reductionism' noted in the main text. (This is the kind of preformationist fantasy behind the narratives and anxieties around genetics in popular culture, from *Jurassic Park* to *Gattaca* to more complicated concerns about GMOs.) The main text here and the argument it makes (following Oyama, Barbieri and others) for a located and situated account and operation of information are in part to be understood against this substantialist notion of information, i.e., as directed towards undermining the assumptions of the determinism of neo-Darwinism and molecular biology.

2. The term 'eventive' is constructed here to parallel 'inventive', whence 'evention', 'eventing', and so on. The neologism is useful as a shorthand way of indicating the dynamic operation of information in and to a system. This serves to indicate, first, how the each-time-event of information is transitive, verb-like and adjectival rather than substantive and noun-like, and, second, a certain proximity and distance to invention. This last point is taken up towards the end of this article.

3. Technical terms that may be required from here (though they are not significant to the present argument and so will not be presented in any detail) include *amino acids* – the base molecules out of which longer chains of certain biologically functioning molecules are made, specifically the following classes of biomolecule: *nucleotides* – chains of amino or nucleic acids (the latter are so called because they are found in cell nuclei), the only relevant ones of which here are DNA and RNA, the former being a famously helixical intertwining of two nucleotides and the basic though complex molecule of the gene; and *polypeptides* – single linear chains of amino acids such as proteins which are to be found throughout the cell and take on a wide variety of functions.

In the neo-Darwinian doctrines of modern (molecular) biology all other molecular activity is subordinated to and *organized by* the supposedly principal nucleotide that is the gene, whence the *genocentrism* that Barbieri critiques through the argument presented in the next few lines of the main text here.

4. Since both genotypes and phenotype and the 'first organic systems' (2003: 145) are constituted by the earlier production of ribonucleo-proteins, Barbieri proposes a *ribotype* theory for the origin of life. Ribotypes are the 'seat of genetic coding' in Barbieri's theory (2003: 156) – which is to say that organic codes and thus organic meaning are the condition for the development of life even at its origin.

5. This statement also characterizes the logic of Derrida's ethics, in particular the

logic of the 'to come' and the quasi-eschatological messiahanism of the promise that have gained increasing prominence in Derrida's work. See, for example, *Specters of Marx* ([1993] 1994: 91). Important differences remain, however: though the following argument in the main text on knowledge is entirely relevant to Virilio, it is only pertinent to Derrida in a severely limited way.

References

Barbieri, Marcello (2003) *The Organic Codes: An Introduction to Semantic Biology.* Cambridge: Cambridge University Press.

Bateson, Gregory ([1972] 2000) *Steps towards an Ecology of the Mind.* Chicago, IL: University of Chicago Press.

Depew, David and Bruce Weber (1995) *Darwinism Evolving: Systems Dynamics and the Genealogy of Natural Selection.* Cambridge, MA: MIT Press.

Derrida, Jacques ([1993] 1994) *Specters of Marx: The State of Debt, the Work of Mourning, and the New International.* New York: Routledge.

Dretske, Fred (1981) *Knowledge and the Flow of Information.* Oxford: Blackwell.

Lash, Scott (2002) *Critique of Information.* London: Sage.

Luhmann, Niklas ([1984] 1995) *Social Systems*, trans. J. Bednarz, Jr. and D. Baecker. Stanford, CA: Stanford University Press.

Lyotard, Jean-François ([1979] 1984) *The Postmodern Condition: A Report on Knowledge*, trans. G. Bennington and B. Massumi. Minneapolis: University of Minnesota Press.

Oyama, Susan (2000) *Evolution's Eye: A Systems View of the Biology–Culture Divide.* Durham, NC: Duke University Press.

Stiegler, Bernard ([1994] 1998) *Technics and Time 1: The Fault of Epimetheus*, trans. R. Beardsworth and G. Collins. Stanford, CA: Stanford University Press.

Virilio, Paul (2003) *Unknown Quantity*, trans. Chris Turner and Jian-Xiang Too. London: Thames and Hudson.

Suhail Malik teaches Postgraduate Fine Art at Goldsmiths College, University of London, and has written several articles on the conceptualization of technoscientific processes and on issues relating to contemporary art practices.

Pharmaceutical Matters
The Invention of Informed Materials

Andrew Barry

Introduction

IN COMPARISON to physics and biology, chemistry appears to be a science lacking in theoretical interest. Unlike physics, it does not claim to be concerned with the investigation of fundamental forces and particles. Unlike biology, it does not concern itself primarily with the properties and dynamics of living materials. Indeed, as Bernadette Bensaude-Vincent and Isabelle Stengers note in their *History of Chemistry* (1996), the discipline is often considered merely a 'service' science. In one common view, although chemistry did once play a leading role in the development of scientific thought, in the 20th century that role seems to have been displaced by other fields. To be sure, chemistry is a large field embracing a huge range of important topics and problems but it apparently no longer possesses the status that it once had in the hierarchy of scientific disciplines: 'chemistry may seem to be a kind of applied physics, whose focus is not on the progress of knowledge but technico-industrial utility' (Bensaude-Vincent and Stengers, 1996: 245). From this perspective, chemistry is doubly uninteresting. First, the direction of its development is determined by purely instrumental considerations. Second, the discipline no longer aspires to address any fundamental questions. At best, contemporary chemistry simply makes it possible for some of the fundamental scientific developments of the 20th century (quantum mechanics and genetics, in particular) to find fields of application. At worst, it remains tied to a naive and outdated ontology of atomism and mechanism. The received view that chemical thought is theoretically limited is not new. In his *Creative Evolution*, Henri Bergson had drawn a sharp contrast between the limitations of physics and chemistry and the philosophical importance of the sciences of life: 'those who are concerned only with the functional [as distinct from the creative] activity of the living being are inclined to believe that physics and chemistry will give us the key to biological processes' (Bergson, 1998: 36; Ansell-Pearson, 1999: 149).

In this article, I make four intersecting arguments, which contest this received view. First, I argue that chemistry is of general interest to social theory not because of its larger theoretical claims or ethical implications, but rather because, as Bensaude-Vincent and Stengers argue, it is an industrial, applied and empirical discipline. Indeed, part of the theoretical interest of chemistry is that it indicates the importance of research which is not primarily guided by theory, but is attentive to the singularity of the case. Second, focusing on a specific case of R&D in pharmaceutical chemistry, I develop Bensaude-Vincent and Stengers's claim that one of the key features of chemical R&D is that it is concerned with the invention of what they term *informed materials*. The article argues that molecules should not be viewed as discrete objects, but as constituted in their relations to complex informational and material environments. Third, drawing on the philosophy of A.N. Whitehead and the sociology of Gabriel Tarde, I suggest how we might make a distinction between the concept of invention, used by Bensaude-Vincent and Stengers, and the concepts of discovery and innovation. While we may or may not agree with Bergson's claim that chemistry does not provide an account of the creative activity of living materials, I argue that chemical R&D is not merely innovative, but is itself creative or inventive, in Tarde's sense of the term. Chemical R&D does not, among other things, discover or synthesize new molecules or new molecular structures but, as Bensaude-Vincent and Stengers argue, it invents informed materials. Fourth, I argue that an important feature of contemporary pharmaceutical chemistry is, to use A.N. Whitehead's terms, the invention of new methods for the invention of such materials. Although the materials produced by chemists have always been informed, the development of contemporary pharmaceutical research has fostered new forms and levels of informational enrichment. My suggestion is that the chemical molecules invented by chemical R&D are now so rich in information that the informational content of invented materials becomes easier to recognize. In part, the conduct of contemporary pharmaceutical R&D is of general interest precisely because it makes the informational content of invented materials more clearly visible. In sociology, as in chemistry, the general interest of the example derives from an attention to its specificity.

Chemistry

In their history, Bensaude-Vincent and Stengers do not deny that there is some truth in the received view of chemistry as merely a service science. But they offer two correctives, both of which suggest a richer account of the history of chemistry. First, they note some of the ways in which chemistry has continued to produce surprising and fundamental results in the 20th century. They point, in particular, to Prigogine's work in far-from-equilibrium physical chemistry and his analysis of self-organizing systems. Contemporary chemistry, in their view, points to the limitations of those approaches that seek to deduce from first principles, but instead recognizes the possibility of learning from the contingent:

> What are the properties of a substance if one is interested only in deducing them without learning? And how does one learn from them if not by painstaking experiments of which they are an integral part or deciphering the temporal configuration of all the processes at work? (Bensaude-Vincent and Stengers, 1996: 264)

In this view, chemistry is not so much a positivist science, but a discipline which points to a new form of empiricism. It produces substances, the properties of which *cannot* be derived from general laws.[1]

Second, and relatedly, their history indicates that the 'technico-industrial utility' of chemistry cannot be understood simply as a process of application. On the one hand, once outside of the laboratory, chemists confront environments or open systems which necessarily do not correspond to the closed environments of the laboratory (Bensaude-Vincent and Stengers, 1996: 249). In these circumstances, the relation between the field of application (factory, urban environment, field) and the laboratory is necessarily one of translation, rather than application or diffusion (Latour, 1988, 1999). On the other hand, in so far as chemistry has played a critical part in the development of new materials, it has also given rise to a different notion of matter. Matter is not merely reshaped mechanically through chemical R&D but is, according to Bensaude-Vincent and Stengers, transformed into *informed material*:

> Instead of imposing a shape on the mass of material, one develops an 'informed material' in the sense that the material structure becomes *richer and richer in information*. Accomplishing this requires a detailed comprehension of the microscopic structure of materials, because it is playing with these molecular, atomic and even subatomic structures that one can *invent* materials adapted to industrial demands. (1996: 206, my emphasis)

Bensaude-Vincent and Stengers's argument raises two immediate questions, to which they do not provide answers in this text. First, what is implied by the idea that such materials are invented? What is at stake in using the term invention in describing what happens to chemical substances in the laboratory rather than, for example, the term discovery? Second, how are we to make sense of the idea that materials can somehow become 'informed' or, as they suggest, 'richer and richer' in information?

Invention

What is an invention? In *The Laws of Imitation*, Gabriel Tarde provides us with a starting point for a social theory of invention. For Tarde, invention was not the opposite of imitation, nor was the relation between invention and imitation analogous to the sociological distinction between agency and structure. Rather, invention involved the novel composition of elements which were themselves imitations: 'all inventions and discoveries are composites of earlier imitations . . . and these composites are, in their turn, destined to become new more complex composites' (Tarde, 2001: 105). In Tarde's ontology there were no fundamental elements from which composites were invented. Even those objects which were often taken to be fundamental – such as chemical atoms and human individuals – were only fundamental from the point of view of specific scientific disciplines.[2]

As a composite, the properties of any invention were not reducible to the elements from which it was composed. At the same time, as Tarde argued, the process of invention provided a direction to history, although one that was neither linear nor predictable. Anticipating the conclusions of more recent economists and sociologists of technology, Tarde recognized that the process of invention was contingent, irreversible and path-dependent. In this way, Tarde conceived of inventions as events, not as mere moments in the progressive evolution of technology or the manifestation of the movement of societies from one form to another: '. . . to establish social science it is not necessary to conceive the evolution of societies . . . with a formula comparable to the type of itinerary planned in advance that the railroad companies propose to and impose on tourists' (1967: 93).

Some of Tarde's comments on invention seem to imply that he viewed individual genius as being of critical importance to the inventive process. Yet his account of invention was not psychological, nor did Tarde have a romantic conception of the individual creator. On the one hand, his account was based on a generalized social psychology of belief and desire, in which the notion of society applied as much to non-human as to human entities (Alliez, 1999). On the other, Tarde recognized that what he termed scientific geniuses (such as Cuvier, Newton and Darwin) mobilized the action of many obscure researchers whose contribution was often ignored (Tarde, 1999: 66). Invention, in Tarde's account, was accomplished not by an individual agent, but by lines of force which came to traverse the individual person. Moreover, for an invention to become irreversible depended on the extent of its subsequent imitation by others.

Tarde's conception of invention provides a corrective to two commonplace ways of conceiving of technological invention in general, and the inventive practice of chemistry in particular. In one view, chemical R&D is driven by social and economic forces. It is a service science, after all. In this way, the products of chemical innovation (such as molecules) become shaped by a social and economic dynamic which was external to them. In Tarde's terms, this form of socio-economic analysis operates with an excessively restrictive conception of society. In effect, the activity of chemical substances is simply rendered inert, excluded from the active realm of the social. In a second view, the chemist works to *discover* new materials. Indeed, the idea that new molecules are discovered is one apparently implied by pharmaceutical chemists themselves who write of research and development as a process of 'drug discovery'.[3] In this account, the fundamental properties of the finite set of chemical elements which make up the periodic table provide a set of given possibilities out of which effective drug molecules can subsequently be synthesized. This view resonates with the 19th-century notion that nature exists as a repository of potential inventions, which are simply there waiting to be realized or discovered by the scientist or engineer (Macleod, 1996). For Tarde, such an account fails to recognize that an atomic element or a molecule is never just an element in isolation. Inevitably, the chemist, in discovering a new molecule, invents a new composite element. Invention leads to the actualization of the virtual, rather than the realization of the possible (Deleuze, 1988: 96–7).

Although Tarde does provide a starting point for a social theory of invention, his own historical analysis fails to recognize the significance of the industrialization of science and engineering that occurred in the late 19th century (Noble, 1977). In this respect, his image of invention was indebted to romantic notions of individual creativity. A.N. Whitehead's later remarks on the history of 19th-century science in *Science and the Modern World* are more suggestive. For Whitehead, 'the greatest invention of the nineteenth century was the invention of the method of invention' (1985: 120). This oft-quoted comment seems remarkable in a book that is primarily concerned with issues in the history and philosophy of science, rather than the sociology and history of technology. Yet it makes perfect sense in the context of Whitehead's philosophical project. In *Science and the Modern World*, Whitehead had little to say about the kinds of problem that traditionally preoccupy philosophers of science such as the relation between theory and evidence or the nature of scientific method. His concerns were metaphysical not epistemological, and at the heart of his philosophy was that 'the ultimate metaphysical principle is the advance from disjunction to conjunction, creating a novel entity other than the entities given in disjunction' (Whitehead, 1978: 21). As for Tarde, Whitehead's was a metaphysics of association. For Whitehead, the 19th-century invention of the method of invention made the production of novel associations a matter of *systematic* research and development. While fields concerned with the invention and investigation of materials such as chemistry and metallurgy have arguably played a merely supportive part in the development of many of the best-known developments in 20th-century scientific theory, from the point of view of the history of invention their role is absolutely critical.[4] To view such fields of science as merely instrumental, or simply driven by an economic logic, would fail to recognize their inventiveness. The notion of informed material, put forward by Bensaude-Vincent and Stengers, points to one way in which such sciences have been inventive, and to one way in which atoms and molecules come to exist, to use Tarde's terms, as 'complex composites'.

Informed Materials

How can we understand the idea that materials can be informed? Two views were commonplace among chemists in the late 19th century. First, in comparison to physics, which sometimes dealt with metaphysical abstractions, chemists prided themselves on the practical craft of their discipline.

In this period, 'chemistry's greatness consisted precisely in its not transcending the facts learned from its practice' (Bensaude-Vincent and Stengers, 1996). Chemistry was a discipline grounded in the controlled environment of the laboratory. Meeting the 'converging interests of academic research and industrial production', the chemistry laboratory both produced new entities and provided the space within which they could reliably be witnessed (Stengers, 1997: 95).

Second, many (although not all) chemists viewed the discipline as a science of atomic elements and molecules. This identity was displayed clearly in the periodic table, a diagram that is still to be found on the walls of the present-day laboratory. Conceived in this way, chemistry appeared to make two assumptions. One was that atoms have given and invariant identities, an assumption which was (partially) undermined with the discovery of radioactivity at the beginning of the 20th century. The second was that chemistry is a science of combinations between these invariant entities. Despite the fact that chemists write of things such as carbon, water and iron all the time, such atoms and molecules are never studied in isolation. The chemist is interested in the fact that the properties of atoms and molecules vary considerably depending on the form and circumstances of their association with others.

For Whitehead, the discipline of chemistry had a particular importance in the exposition of his philosophy of organic mechanism. For Whitehead recognized that the image of matter as being composed of distinct atoms and molecules had come to inform contemporary understandings of reality. In this commonplace view, a molecule is thought of as something like a stone – a kind of stuff 'which retained its self-identity and its essential attributes in any portion of time' (Whitehead, 1978: 78; Stengers, 2002). Whereas Bergson wished to distance himself from what he viewed as the limitations of chemical thought, Whitehead's own criticism of this commonplace view drew some inspiration from chemistry. In his account, however, the identities of atoms and molecules were not distinct, nor were they invariant. Rather than starting out from the first assumption (the invariability of atoms and molecules), Whitehead began from the second (the variability of their associations). Viewing chemistry as a science of associations or relations, Whitehead argued that a molecule should be considered an historical rather than a physical entity. In his view, a molecule should not be understood as a table or rock, but rather as an event: 'a molecule is a historic route of actual occasions; and such a route is an event' (Whitehead, 1978: 80). Seen in these terms the endurance of a molecule through time cannot be taken for granted. Molecules certainly endure, but it cannot be assumed that they remain the same: 'physical endurance is the process of continuously inheriting a certain identity of character transmitted throughout an historical route of events' (Whitehead, 1985: 136).

Chemistry should not be understood then as a science of combinations between given elements that are nonetheless to be considered distinct and immutable. Rather, the identity and properties of atoms and molecules are transformed through their changing associations. The properties of a hydrogen atom bound within a water molecule are different from the properties of a hydrogen atom bound within a hydrogen molecule. The properties of a water molecule are quite different at temperatures above and below 0°C. The properties of a metal vary considerably depending on whether it contains trace impurities of other elements. In displacing the notion of the object by the notion of the actual occasion or actual entity, Whitehead suggested a different account of atoms and molecules. For Whitehead, actual entities, including molecules, are not bounded at all, but are extended into other entities, while folding elements of other entities inside them. As became clear with the development of quantum chemistry, apparently distinct atoms and molecules entered into the internal constitution of others through their association. This recognition was a central part of Whitehead's metaphysics: '[An] actual entity is present in other actual entities . . . The philosophy of organism is mainly devoted to the task of making clear the notion of "being present in another entity"' (1978: 50, see also Deleuze, 1993: 78; Halewood, 2003).

For chemists, the fact that molecules have changing properties depending on their associations is an everyday reality. The molecule that is isolated and purified in the laboratory will not have the same properties as it has in the field, the city street or the body (Barry, 2001: 153–74). The challenge, for the chemist, is to multiply the relations between different forms of existence of a molecule both inside and outside the laboratory (Latour, 1999: 113–14). It is impossible to establish an identity between the molecule in the laboratory and a molecule elsewhere, but it may be possible to establish a relation of translation. The problem is particularly difficult to address in thinking about the properties of drugs. Bensaude-Vincent and Isabelle Stengers note the challenge faced by chemists engaged in pharmaceutical research:

> The pharmacological chemist can certainly pursue the dream of an *a priori* conception of molecules to be synthesized for their pharmaceutical properties, but it is still the case that 60 to 70 per cent of medicines today are of natural origin . . . From this field the chemist takes the active molecules, which he isolates, purifies and copies, and modifies at leisure. But it is also 'on the field – on the ailing body' – that medicine designed in a laboratory must operate. Humanity delegates active chemical substances to act not in the aseptic space of a laboratory but in a living labyrinth whose topology varies in time, where partial and circumstantial causalities are so intertwined that they escape any *a priori* intelligibility. (1996: 263)

Bensaude-Vincent and Stengers pose the problem of the relation between the 'aseptic space of a laboratory' and the 'living labyrinth' of the body as an ontological one. Molecules necessarily do have different identities and effects in the laboratory and the body. But, for pharmaceutical research, the gap between the laboratory and the body is equally economic, regulatory and legal. Although pharmaceutical companies may be able to identify potential drug molecules through a variety of methods, there is no guarantee that active molecules will work effectively and safely as drugs in living bodies.[5] During development many active molecules fail, whether because they are poorly absorbed or metabolized, or are subsequently shown to have toxic effects. Moreover, in the context of the growing concern of consumers, regulators have become more cautious about drug approvals and 'increasing post-marketing surveillance has led to an increasing number of withdrawals'.[6] The withdrawal of Bayer's Baycol™ is a well-known recent example.[7] In these circumstances, research and development costs have escalated. Pfizer, for example, the world's largest drugs company, has warned that its $5bn annual research budget will yield only about two major new drugs per year. The average pre-clinical trial development cost of new chemical entities is said to be $30m per molecule. Perhaps 90 percent of such molecules fail such trials. The cost of generating a single approved medicine is claimed to be over $600 million.

For pharmaceutical companies the costs of clinical trials and the even greater costs of withdrawing drugs after they have been marketed pose a clear problem: how is it possible to maximize the chances that a drug will be both effective and safe prior to the conduct of such trials and, thereby, to increase the productivity of pharmaceutical R&D? How can reliable relations be established between the 'aseptic space of a laboratory' and the 'living labyrinth' of the body without the presence of real bodies? In brief, how can innovation be speeded up?[8]

Bensaude-Vincent and Stengers indicate one solution to the problem. Pharmaceutical R&D can be directed to the extraction and purification of active molecules from naturally occurring substances. This practice can give rise to a series of legal and ethical questions concerning the ownership of intellectual property, for example, regarding indigenous knowledge of the medicinal properties of plants, or the ownership of viruses which are present in particular populations (Pottage, 1998). Such an approach is of continuing importance, but it is only one possible research strategy open to pharmaceutical companies. A more general understanding of pharmaceutical R&D is suggested by the notion of informed material.

One way of understanding the idea that a material entity (such as a potential drug molecule) could be informed or 'rich in information' would be to say that the material *embodies* information. In this view, the design process builds information into the structure of the molecule. But this view would not make sense if we understood the molecule to be simply a discrete and bounded entity. For if molecules were simply discrete entities, how could one then distinguish between a molecule which embodies little information and the 'same' molecule with the same structure of elements that embodies a great deal of information? In Whitehead's and Stengers's terms it is possible to give a different and more precise meaning to the idea of a material object being rich in information. This would acknowledge that material objects (such as molecules) exist in an informational and material environment, yet this environment cannot, as Whitehead argued, be considered as simply external to the object. An environment of informational and material entities *enters into* the constitution of an entity such as a molecule. Nor can this environment be perceived from a viewpoint which is external to it. The perception of an entity (such as a molecule) is part of its informational material environment (Whitehead, 1985: 87; Fraser, 2002).

Thus defined, the notion of an informed material makes sense of what pharmaceutical actually do. Pharmaceutical companies do not produce bare molecules – structures of carbon, hydrogen, oxygen and other elements – isolated from their environments. Rather, they produce a multitude of informed molecules, including multiple informational and material forms of the same molecule. Pharmaceutical companies do not just sell information, nor do they just sell material objects (drug molecules). The molecules produced by pharmaceutical companies are more or less purified, but they are also enhanced and enriched through laboratory practice. The molecules produced by a pharmaceutical company are already part of a rich informational material environment, even before they are consumed. This environment includes, for example, data about potency, metabolism and toxicity and information regarding the intellectual property rights associated with different molecules. In this way, pharmaceutical laboratories have similarities to other laboratories. As Karin Knorr-Cetina argues, laboratories: 'invent and recreate . . . objects from scratch . . . creat[ing] new configurations of objects that they match with an appropriately altered social order' (1999: 44).

Drug Discovery

Consider the case of a medium-sized pharmaceutical company called ArQule which, towards the end of the 1990s began to transform itself into a 'drug discovery company' – a company oriented toward the development of new chemical entities. Although ArQule was unusual in some respects, its approach to drug discovery is indicative of broader shifts in the conduct of contemporary pharmaceutical R&D. These centred on the introduction of new technologies, including high-throughput screening, combinatorial chemistry, genomics, and computer modelling (Bailey and Brown, 2001).[9] In this way, elements of the drug discovery process, which had hitherto been based on craft laboratory skills, became increasingly industrialized (Augen, 2002; Handen, 2002). At the same time, the introduction of new technologies involved alliances between companies working in distinct areas of technology, and also the formation of so-called virtual pharmaceutical companies which managed such alliances (Cavalla, 2003: 267).

ArQule made its name as a pioneering company in combinatorial chemistry, a set of techniques that made it possible to produce a huge number of potential drug molecules cheaply and quickly.[10] Instead of being the product of specific synthetic pathways of the kind associated with traditional synthetic organic chemistry, combinatorial chemistry enabled new molecules to be mass produced (Hird, 2000; Thomas, 2000: 69–88).

Synthetic chemistry: A + B + C -> AB + C -> ABC (an individual compound)

Combinatorial chemistry: $A_n + B_n + C$ -> $A_nB_n + C_n$ -> $A_nB_nC_n$ (combinatorial library of 10,000–1,000,000 compounds)

For the traditional organic chemist the problem was to find the most efficient way of synthesizing a given molecular compound (ABC) from a finite set of building blocks of existing compounds (A, B, C, D . . .) which were either readily available in the laboratory or could be purchased from chemical suppliers. Indeed, the discovery of solutions to particular synthetic problems was central to the field of organic chemistry, as it was once taught in university courses. In the laboratory, organic chemists had to deal with all the difficulties of translating formal solutions to synthetic problems into practice. As Bensaude-Vincent and Stengers explain:

> organic chemistry texts usually present the classic, conventional reaction chains. But to the student or researcher falls the problem of directing the actors in a play, so to speak, and creating the situations they need to achieve the desired goal. (1996: 159)

By contrast, combinatorial chemistry performs synthesis through mass production. Through combinatorial chemistry a large number of different but chemically similar building blocks (A_a, A_b . . . A_n) can be reacted with sets of other building blocks (B_a, B_b . . . B_n) and (C_a, C_b . . . C_n) to produce huge numbers of synthetic compounds. In this way, molecules come to exist not as the product of individual synthetic pathways, as was previously the case, but in conjunction with a multitude of other molecules produced through combinatorial pathways. Physically, molecules produced through such techniques are dissolved in standard solutions and stored, for example, in arrays of test-tubes. These arrays collectively form what in the industry are termed *libraries* of compounds (Beno and Mason, 2001). The metaphor of a library of molecules is appropriate because not only do such mass-produced molecules have a material existence, but they are also held in an informational form in catalogues and databases. Without further research, individual molecules produced through combinatorial chemistry have little commercial value. In practice, combinatorial chemistry companies, such as ArQule, sold whole libraries of molecules to those larger pharmaceutical companies that had the resources to investigate and exploit them.

But although combinatorial chemistry, in conjunction with high-throughput screening techniques, reduced the costs of producing and analysing the properties of new molecules, it did not solve the problem of how to determine whether they would work in living bodies. According to industry reports, combinatorial chemistry companies faced the problem that the new technology was not yielding the kinds of dramatic improvements in the productivity and efficiency of drug discovery that had been anticipated by investors and partners. The danger was that ArQule would end up simply providing the bulk material for drug development, but not playing any significant role in the subsequent informational enrichment of its product. It would not be able to engage in either what researchers term 'lead generation' (developing a set of molecules which have the potential to become drugs) or 'lead optimization' (refining this set). In these circumstances, ArQule's strategy was to reinvent itself as a 'drug discovery company' and, at the same time, to attempt to create a new form of informed material. Value could be realized by enriching molecules with information.

In broad terms, ArQule's attempt to do this had two elements. One was to integrate elements of the existing 'drug discovery process'. To work as a drug, a molecule did not merely have to be potent; it also had to be absorbed by (and eliminated from) the body, it had to be non-toxic, and metabolized neither too slowly nor too quickly. Traditionally, major pharmaceutical companies had performed tests for these properties in sequence. First, potential drug candidates were tested for potency against specific targets, then the other properties of those molecules that were likely to be potent were investigated. ArQule's aim was to perform them in parallel, thereby dramatically reducing the time taken to optimize the design of a potential drug molecule:

> In the traditional drug discovery process, physico-chemical properties, selectivity, potency and ADMET (absorption, distribution, metabolism, elimination, toxicity) parameters are evaluated in a sequential manner, extending the time required to identify a lead candidate and increasing costs. Key information provided by ADMET profiling is historically obtained *at the end of the discovery process*. Adverse results at this step can eliminate compounds that have already progressed for many years, at a substantial cost (Arqule, 2001) . . . [Instead, the] sequential process with late failures must be replaced by a multi-parameter filer at every stage of the drug discovery process. (Hill, 2001)

This strategy was called Parallel Track™ drug discovery: the trade mark is an indicator that this method of invention itself had a market value and public visibility. As a brand, ArQule did not address itself to consumers (Blackett and Robins, 2001), but to the network of potential investors, collaborators and researchers necessary to maintain an innovative company.

The second element of ArQule's approach to the problem of drug discovery involved a proliferation of the forms of existence of molecules. Molecules increasingly existed in ArQule not merely as material and informational objects in laboratories and libraries, but also as the objects of computer modelling. To be sure, computational methods had already established a place in the drug discovery process, however, this place had been a limited one:

> [In drug discovery a] team must come up with a drug which will interact with a novel target for therapeutic intervention in an important disease. The team will have access to data on related targets and existing drugs which interact

with them. They may have a crystal structure of a target protein. And using a library of computational tools, with their inherent sets of chemical rules, the team can make an informed assessment regarding the shape of molecules that might interact with the target. In modern companies, they will then be able to enumerate a focused library of possible actives using these methods. But this is where their 'simulation' ends. (Beresford et al., 2002)

ArQule's approach was to extend the use of computer models to the simulation of ADMET. Through computer models, the libraries of molecules generated through combinatorial chemistry could be subject to what pharmaceutical researchers called *virtual screening* (Manly et al., 2001). In this way, it would become much easier, and cheaper, to deal with the size of library generated by combinatorial chemistry. In principle, huge libraries of molecules could be enriched through computer modelling, reducing the need for costly laboratory experiments. In practice, however, the development of computer models which might be of use to the laboratory chemist is far from straightforward. Models themselves can be derived, in part, from general quantum mechanical principles. But, as Bensaude-Vincent and Stengers argue, chemistry can rarely rely on general principles, perhaps particularly in the case of the pharmaceutical industry. Necessarily, the development of computer models relies on data derived from earlier laboratory and clinical trials on molecules that may be more or less different from the molecules that the chemist is interested in. However sound the theoretical bases of models are, their reliability depends on the quality and breadth of the data sets on which specific calculations are based.

In discussions between chemists, the term *chemical space* has particular importance. Why is this term so significant? One reason is that it provides a way of thinking about the distance between the properties of the molecules they are interested in and the properties of the molecules that have been used to derive the models. The quality of the models depends on the volume of chemical space they are able to operate within with some degree of reliability. As one team of chemical modellers explained: 'Any primordial models in the past were invariably poor in their predictability because they were based on a very small data set of tens of compounds' (Beresford et al., 2002). In this way, the concept of chemical space is both important, but difficult to operationalize. It is not a Newtonian space, governed by particular coordinate axes which exist independently of the entities which exist within the space. Rather, chemical space is a relational space, the coordinates of which are governed by the particular medical chemical process under investigation. Two different molecules which exist in close proximity to each other in relation to one specific process, for example, may be distant from each other when viewed in relation to a different process. Different pharmaceutical companies, research teams or projects may temporarily occupy different regions of chemical space. But, at the same time, they are likely to conceive of the structure of chemical space in quite different ways.

While computer modelling can be used to select molecules from the libraries generated through combinatorial chemistry, modelling also generates and tests molecules that may not necessarily have any material existence at all. Molecules can be synthesized on screen – even more easily than through combinatorial chemistry. As well as combinatorial libraries, it is now possible for pharmaceutical companies to hold virtual libraries of molecules that have never been synthesized. However, it should not be thought that such computational experiments are necessarily less real than those tested in a traditional laboratory. For some pharmaceutical researchers and managers, all techniques are viewed more or less instrumentally in terms of how quickly and efficiently they yield molecules with a potential to become drug molecules. Others point out that computational experiments are closer to external reality than traditional laboratory experiments as they are likely to be based on data derived from trials on living bodies, whereas laboratory experiments will be conducted in standard solutions.[11] Linguistically, researchers establish equivalence between experiments conducted through computer models by computational chemists and experiments conducted using chemical materials by laboratory chemists. The former experiments are *in silico*, the latter are *in vivo* or *in vitro* (Leach and Hann, 2000). For the chemist it would not make sense to say that the experiment that takes place through computer model is simply a representation of the kind of experiment traditionally carried by the organic chemist or biologist in the laboratory. *In silico*, *in vivo* and *in vitro* experiments are all considered as distinct events that constitute their own objects, relations and forms of measurement, and their own strengths and weaknesses. The problem for a pharmaceutical research group is to translate between the different forms of experiment and different forms of existence of molecules, so that they enrich each other. In practice, this translation is likely to be difficult.

ArQule did not produce molecules that could be sold directly to consumers. The company did not have the resources to invest in expensive clinical trials, nor the political and legal expertise with which to manage its relations with the regulatory authorities, nor the infrastructure required for marketing and distribution. In this context, measurements of the properties of molecules, in their various material and immaterial forms, are critical to the formation of the market for potential drug molecules for companies such as ArQule (cf. Callon et al., 2002: 198–9). For along with other pharmaceutical companies, ArQule had to develop new materials that were sufficiently rich in information that they could provide both the basis for claims to intellectual property, and that would also be likely candidates for further development. Potential purchasers of ArQule's research did not purchase molecules, but molecules the properties of which had, in various forms, been measured, and that were, thereby, uncertainly predictive of their clinical existence. The economics of the pharmaceutical industry revolve around an extraordinary level of investment in measuring equipment – including computer modelling technology, laboratory tests and clinical trials – which produce uncertain results.

Thus specific molecules exist in the informational and material environment of the laboratory. But they also exist in a legal and economic environment of other molecules developed by other companies. Necessarily, in formulating research strategies chemists take into account the existence of prior patents.[12] This information, updated daily, is available on commercial databases. The importance of this informational environment accounts for a second sense in which chemists use the term chemical space which refers to the distance between patented drug molecules and the set of molecules they are investigating. In the context of this information, chemists may seek to buy into the legal-chemical space owned by other companies through collaboration. But they may also try to develop molecules that exist just outside of the space defined by a patent,[13] or colonize unexplored volumes of chemical space, or attempt to re-design drug molecules which have been patented, but which have failed clinical trials.[14] Moreover, in developing computer models, chemists make use of publicly available data on patented molecules. In these ways, intellectual property law should not be considered as simply a part of the external environment within which pharmaceutical companies operate and drug molecules are developed. In a number of different ways, information about existing patents enters into the life of molecules, even during the earliest stages of their development. The molecules produced by the pharmaceutical laboratory are rich in information about their (global) legal and economic, as well as their chemical relations to other molecules. The pharmaceutical laboratory is not a closed system, but a space which itself includes its external legal and economic environment (cf. Mitchell, 2002: 303; Strathern, 2002).

Conclusion: Chemical Invention

Does it make sense to describe pharmaceutical R&D as an inventive practice, rather than merely a practice of discovery? Are the kinds of entities produced through pharmaceutical companies novel? Certainly, the molecules developed by pharmaceutical R&D do not exist 'in nature'. But nor can they be designed simply on the basis of fundamental chemical principles. Nor are they merely structures of atomic elements that always had the potential to be discovered or realized. As we have seen, the development of new drugs involves the multiplication of forms of existence of molecules. But, at the same time, the multiplication of forms of existence of molecules is associated with their progressive informational enrichment.

Thus, the kinds of entities that are produced by pharmaceutical R&D are not simply bare molecules. Rather, they can be understood as 'societies' of different elements, as long as we understand that societies are associations of non-human as well as human entities. The idea of chemical space, which is used so frequently by pharmaceutical chemists, conveys precisely the way in which chemists understand that molecules are societies, in Tarde's sense of the term. As I have argued, the kinds of societies produced by pharmaceutical R&D take specific historical forms. The molecules produced in the contemporary pharmaceutical laboratory certainly are more or less purified as chemicals, but they are also enriched in new ways. They are part of increasingly dense, spatially extended and changing informational and material environments formed not just through laboratory syntheses and tests, but through virtual libraries, computational models and databases. The notion of 'informed materials', introduced by Bensaude-Vincent and Stengers, describes such novel entities very well.[15]

For Whitehead it was a mistake to imagine that material objects (such as molecules) ever had a concrete existence. Rather than imagine that there are concrete material objects to which social meanings and uses are then added, he argued that objects themselves take historical forms. Whitehead himself was preoccupied by the problem of how, despite the historicity of things, things didn't change that much. Things endured. One of the key assumptions of chemistry is, of course, endurance. Atoms and molecules are never exactly the same as they were before, depending on their changing environments, but they also have an amazing capacity for endurance. Within the drug discovery process, the forms of existence of molecules proliferate. Molecules have characteristics and properties depending on their existence in different informational material forms (in laboratory tests, clinical trials, computer models, patent databases, etc.). But this does not mean that the identities of molecules are fluid. On the contrary, pharmaceutical research can only proceed on the basis that molecules actually endure across different sites, through different parts of the laboratory, throughout their life as products. In the pharmaceutical laboratory, the generation of enduring novel entities depends upon the multiplication of different forms of informed material.

Acknowledgements

My thanks to Georgie Born, Monica Greco, Mariam Fraser, Sarah Kember, Celia Lury, Mick Halewood and the referees for *Theory, Culture & Society* for their comments on an earlier draft of this article and to Alan Blackwell of Crucible for his support and collaboration. Thanks also to ArQule and Camitro and, in particular, to Mike Tarbit, Matt Segall, Mark Ashwell and Steve Gallion for their support, interest and comments.

Notes

1. In the sense given to the idea of empiricism by Whitehead and taken up by Deleuze: 'the abstract does not explain, but must itself be explained; and the aim is not to rediscover the eternal or the universal, but to find the conditions under which something new is produced' (Deleuze and Parnet, 1987: vii).

2. 'The final elements that every science ends up with – the social individual, the living cell, the chemical atom – are final only with respect to their particular science. They are themselves composite' (Tarde, 1999: 36, cited in Alliez, 1999: 10).

3. One of the leading trade journals of the industry is called *Drug Discovery Today*, and chemists speak of pharmaceutical companies as 'drug discovery companies'. The frequent use of the term discovery does not mean, however, that chemists understand the term literally.

4. Whitehead noted the critical importance of metallurgy to the development of physics in the early 20th century:

The reason why we are on a higher imaginative level is not because we have finer imagination, but because we have better instruments. In science, the most important thing that has happened over the last forty years is the advance of instrumental design. This advance is partly due to a few men of genius such as Michelson and the German opticians. It is also due to the progress of technological processes of manufacture, particularly in the region of metallurgy. (1985: 143)

5. I leave aside here the critical question of the politics of clinical trials and the relations between pharmaceutical companies and regulatory agencies. For further discussion of these issues see Abraham (1995).

6. *The Financial Times*, nd, 2001.

7. In August 2001 Bayer voluntarily withdrew Baycol from the US market because of reports of sometimes fatal rhabdomyolysis, a severe muscle adverse reaction (Food and Drug Administration, 2001).

8. Macdonald and Smith (2001: 947) give an indication of the pressures placed on pharmaceutical R&D for increased productivity in the late 1990s:

> In 1998 GlaxoWellcome embarked upon a new enzyme-inhibitor programme . . . [featuring] an aggressive timeframe of seven years, from the start of medicinal chemistry through to drug launch. This period, dominated as it was by the constraints of the clinical programme [i.e. of testing on human patients], translated into a lead-optimization phase [i.e. the period in which likely potential drug molecules are identified prior to clinical trial] of no more than 12 months.

9. While the ethical and political implications of genomics have been a key focus for research in the social sciences, the development of genomics has seldom been placed in the context of other related trends in research and development. At the same time, elements of the drug discovery process, which had hitherto been based on craft laboratory skills, became increasingly industrialized (Augen, 2002; Handen, 2002).

10. Later commentators indicate that combinatorial chemistry became, for a period, an industrial fashion, just as genomics was later in the 1990s:

> the launch of combinatorial chemistry onto an unsuspecting pharmaceutical industry in the early 1990s resulted in several frantic efforts as companies tried to maintain a competitive edge through the generation and screening of compounds in unprecedented numbers and at an unprecedented rate. (Everett et al., 2001: 779)

The importance of speed in the commercial development of chemistry is not new. Synthetic chemists have often been concerned with the question of the speed and productivity of reactions and the whole field of catalysis derives from this concern.

11. In a pharmaceutical laboratory, potential drug molecules will generally be tested in solution. The solutions used by different laboratories need to take standard forms in order for results of different experiments to be comparable (Cambrosio and Keating, 1995: 82). Such standard solutions can never correspond to the more complex and variable conditions found in a living body. For examples of the

presentation of results of computational experiments see http://www.documentarea.com/qsar/a_beresford2002.pdf

12. See, for example, *The Investigational Drugs Database,* which

> is a daily-updated, enterprise-wide competitor intelligence and R&D monitoring service. It provides validated, integrated and evaluated information on all aspects of drug development, from first patent application to launch or discontinuation. Subscribers include most major pharmaceutical and biotechnology companies the world over. In addition, more and more companies servicing the pharmaceutical and biotechnology sector are subscribing. Chemical companies, CROs, consultants and media providers find the IDdb3 invaluable in locating lucrative new business partners. (http://www.iddb3.com/cds/solutions.htm)

13. A patent is likely not to apply to one molecule but to a set of molecules with similar structure (the 'scaffold') and similar biological activity.

14. This strategy is termed 'drug rescue' by researchers. On the relation between the dynamics of innovation and the occupation of technological space more broadly, see Barry (1999/2000).

15. Scott Lash argues that information should be understood as more than merely a collection of signals or data:

> The constant bombardment by signals, the ads of consumer culture and the like does not constitute information. It is chaos, noise. It only becomes information when meaning is attached to it. Information only happens at the interface of the sense-maker and his/her environment. (2002: 18)

Lash's analysis of information has parallels with my analysis of chemical material. The molecules produced through the industrial process of combinatorial chemistry can be thought of as material forms of noise that need to be filtered before they become useful. Individual molecules only become progressively informed in the assemblage of pharmaceutical research.

References

Abraham, J. (1995) *Science, Politics and the Pharmaceutical Industry: Controversy and Bias in Drug Regulation.* London: UCL Press.

Alliez, E. (1999) 'Tarde et le problème de la constitution', introduction to G. Tarde *Monadologie et Sociologie.* Paris: Institut Synthélabo.

Ansell-Pearson, K. (1999) *Germinal Life: The Difference and Repetition of Deleuze.* London: Routledge.

ArQule (2001) ArQule corporate website, http://www.arqule.com

Augen, J. (2002) 'The Evolving Role of Information Technology in the Drug Discovery Process', *Drug Discovery Today* 7(5): 315–23.

Bailey, D. and D. Brown (2001) 'High-Throughput Chemistry and Structure-Based Design: Survival of the Smartest', *Drug Discovery Today* 6(2): 57–9.

Barry, A. (1999/2000) 'Invention and Inertia', *Cambridge Anthropology* 21(3): 62–70.

Barry, A. (2001) *Political Machines: Governing a Technological Society.* London: Athlone Press.

Barry, A. and D. Slater (2002) 'The Technological Economy', *Economy and Society* 31(2): 175–93.

Beno, B. and J. Mason (2001) 'The Design of Combinatorial Libraries Using Properties and 3D Pharmacophore Fingerprints', *Drug Discovery Today* 6(5): 251–8.

Bensaude-Vincent, B. and I. Stengers (1996) *A History of Chemistry*. Cambridge, MA: Harvard University Press.

Beresford, A., H. Selick and M. Tarbit (2002) 'The Emerging Importance of Predictive ADME Simulation in Drug Discovery', *Drug Discovery Today* 7: 109–16.

Bergson, H. (1998) *Creative Evolution*. Mineola, NY: Dover Publications.

Blackett, T. and R. Robins (2001) *Brand Medicine: The Role of Branding in the Pharmaceutical Industry*. Basingstoke: Palgrave.

Callon, M., C. Méadel and V. Rabeharisoa (2002) 'The Economy of Qualities', *Economy and Society* 31(2): 194–217.

Cambrosio, A. and P. Keating (1995) *Exquisite Specificity: The Monoclonal Antibody Revolution*. New York: Oxford University Press.

Cavalla, D. (2003) 'The Extended Pharmaceutical Enterprise', *Drug Discovery Today* 8(6): 267–74.

Deleuze, G. (1988) *Bergsonism*. New York: Zone.

Deleuze, G. (1993) *The Fold: Leibniz and the Baroque*. London: Athlone.

Deleuze, G. and C. Parnet (1987) *Dialogues*. London: Athlone Press.

Everett, J., M. Gardner, F. Pullen, G.F. Smith, M. Snarey and N. Terrett (2001) 'The Application of Non-Combinatorial Chemistry to Lead Discovery', *Drug Discovery Today* 6(15): 779–85.

Food and Drug Administration, Center for Drug Evaluation and Research (2001) 'Baycol Information', http://www.fda.gov/cder/dru/infopage/baycol/default.htm

Fraser, M. (2002) 'What Is the Matter of Feminist Criticism?', *Economy and Society* 31(4): 606–25.

Halewood, M. (2003) 'Materiality and Subjectivity in the Work of A.N. Whitehead and Gilles Deleuze: Developing a Non-Essentialist Ontology for Social Theory', unpublished PhD thesis, University of London.

Handen, J. (2002) 'The Industrialization of Drug Discovery', *Drug Discovery Today* 7(2): 83–5.

Hill, S. (2001) 'Biologically Relevant Chemistry', *Drug Discovery World* Spring: 129–30.

Hird, N. (2000) 'Isn't Combinatorial Chemistry just Chemistry?', *Drug Discovery Today* 5: 307–8.

Knorr-Cetina, K. (1999) *Epistemic Cultures: How the Sciences Make Knowledge*. Cambridge, MA: Harvard University Press.

Lash, S. (2002) *Critique of Information*. London: Sage.

Latour, B. (1988) *The Pasteurization of France*. Cambridge, MA: Harvard University Press.

Latour, B. (1999) *Pandora's Hope: Essays on the Reality of Science Studies*. Cambridge, MA: Harvard University Press.

Leach, A. and M. Hann (2000) 'The *In Silico* World of Virtual Libraries', *Drug Discovery Today* 5: 326–36.

Levere, T. (2001) *Transforming Matter: A History of Chemistry from Alchemy to the Buckyball*. Baltimore, MD: Johns Hopkins University Press.

Macdonald, S. and P. Smith (2001) 'Lead Optimization in 12 Months? True Confessions of a Chemistry Team', *Drug Discovery Today* 6(18): 947–53.

Macleod, C. (1996) 'Concepts of Invention and the Patent Controversy in Victorian Britain', in R. Fox (ed.) *Technological Change: Methods and Themes in the History of Technology*. Amsterdam: Harwood Academic.

Manly, C.J., S. Louise-May and J. Hammer (2001) 'The Impact of Informatics and Computational Chemistry on Synthesis and Screening', *Drug Discovery Today* 6(21): 1101–10.

Mitchell, T. (2002) *Rule of Experts: Egypt, Techno-Politics, Modernity*. Berkeley: University of California Press.

Noble, D. (1977) *America by Design: Science, Technology and the Rise of Corporate Capitalism*. New York: Oxford University Press.

Pottage, A. (1998) 'The Inscription of Life in Law: Genes, Patents and Bio-Politics', *Modern Law Review* 61(5): 740–65.

Shapin, S. and S. Schaffer (1985) *Leviathan and the Air-Pump: Hobbes, Boyle and the Experimental Life*. Princeton, NJ: Princeton University Press.

Stengers, I. (1997) *Power and Invention: Situating Science*. Minneapolis: Minnesota University Press.

Stengers, I. (2002) *Penser avec Whitehead: Une libre et sauvage création de concepts*. Paris: Seuil.

Strathern, M. (2002) 'Externalities in Comparative Guise', *Economy and Society* 31(2): 250–67.

Tarde, G. (1967) *On Communication and Social Influence*. Chicago, IL: Chicago University Press.

Tarde, G. (1999) *Monadologie et Sociologie*. Paris: Institut Synthélabo.

Tarde, G. (2001) *Les Lois de l'imitation*. Paris: Seuil.

Thomas, G. (2000) *Medicinal Chemistry: An Introduction*. New York: John Wiley & Sons, Ltd.

Whitehead, A.N. (1978) *Process and Reality*. New York: Free Press.

Whitehead, A.N. (1985) *Science and the Modern World*. London: Free Association Books.

Andrew Barry is Reader in Sociology at Goldsmiths College, University of London. He is the author of *Political Machines: Governing a Technological Society* (Athlone, 2001) and co-editor of *Foucault and Political Reason* (UCL Press, 1996) and *The Technological Economy* (Routledge, 2004).

The Performativity of Code
Software and Cultures of Circulation

Adrian Mackenzie

Introduction

IN A recent novel, *Distraction*, Bruce Sterling (1998) describes a future scenario in which information networks have turned bad. By 2044, China has flooded the world's computer networks with pirated copies of commercial software. The US economy, long supported by monopolistic intellectual property arrangements, has consequently collapsed. A populous underclass of unemployed technicians, programmers and engineers, calling themselves Moderators, roam through splintered urban and rural zones in bands, harvesting and recycling technological junk and waste products, extracting energy, discarded components and materials and converting them into tools, energy and materials for their own use. At the centre of post-consumer nomadic Moderator life stands an important infrastructural component: the servers. The Moderators' servers monitor, collate and record the status of members of the community. Individual status fluctuates in real-time in response to continual polling by the servers of the community's opinion of individual contributions to the life of the collective.

Sterling's scenario is not too remote from some contemporary realities. Although software developers do not in any large measure regard themselves as a disenfranchised nomadic underclass, and property rights have not collapsed (in most domains), software cultures increasingly display the kinds of complex collective orderings that Sterling describes. Important sectors of software production are in transnational migration – between the UK/Europe/the USA and Asia – and in trans-institutional movement between corporations, universities and loose aggregates of paid and unpaid workers. Software workers move from Asia to Europe and the USA under special immigration quotas at the same time as European and American companies are outsourcing their software development to software houses in India or the Caribbean. At the same time, self-organized software projects relying on network servers generate large-scale code objects such as operating systems and web servers, which leading computing enterprises such as IBM and Apple take up, sponsor and promote.[1] Programmers such as Linus Torvalds, Larry Wall, Alan Cox, and Richard Stallman reach celebrity status within programming cultures, and minor stardom in the mass media.

Although there has been wide acknowledgement of the mobility, dynamism and operationality associated with information networks (Castells, 2001; Poster, 2001; Lash, 2002), understanding the cultural specificity of software or code objects remains difficult. That specificity is problematic: software has been heavily commodified since the early 1980s when personal computers first began to change from hobbyist devices to office computers running spreadsheets such as *Visicalc* and word processing programs such *WordStar* (Lohr, 2002). Yet software resists commodification in various ways – the tools needed to develop software are widely available and proprietary claims over algorithms are hard to defend. Software often remains invisible because it is infrastructural (Bowker and Star, 1999) and distributed through many different channels. Software production and consumption are also the object of intensely embodied identifications and personal styles (Perl programmers vs. Python programmers, Java programmers vs. C++ coding). It is divided into platform-specific subcultures (Unix programmers vs. Windows programmers). It is at once thoroughly pervaded by interlinked global standards and conventions (such as communication protocols), and at the same time is anarchically polymorphic and mutable. New conventions constantly compete with existing standards. Numerous debates, identities, forms of commodification, capitalization, and regulation swirl around software. Establishing a critical yet synoptic cultural viewpoint on software and code as operational objects remains difficult.[2]

This article investigates some contemporary migrations, translocations and twists in the technoscape[3] associated with one highly complex and polymorphous code object, the Linux kernel, the core component of the GNU/Linux operating system. Linux is often cited as the primary example of free software or 'Open Source Software'[4] (itself a term coined by commercial software producers and computer book publishers in 1998 to dispel the more financially disturbing connotations of 'free' software during the dotcom boom). Linux has made the cover of *Time*, it has frequently been discussed in editorials, received the Golden Nica prize of Ars Electronica (Ars Electronica, 1999), it has generated substantial speculative activity on financial markets during the late 1990s (Taylor, 1999). It has enrolled tens of thousands of programmers and software developers in (mostly) unpaid software development projects. At the same time, Linux, and the other major open source success story, Apache, a web server, have been fairly quickly slotted into corporate software production at companies such as IBM, Compaq, Hewlett Packard, Apple, and Sun Microsystems. In the meantime, open source software has generated huge quantities of meta-commentary about software (of which this article forms a part), and challenged the black-boxing of commercial software in important domains.

Technically, the Linux kernel, dating from late 1991, is not an original or totally new thing. As is well known, it explicitly *clones* another older mainstream operating system, Unix, itself dating from the late 1960s. Organizationally too, Linux is a highly centralized project in many ways. The very term *kernel*, or core, implies a degree of centring that more radical technical architectures have dispensed with. Even after 10 years, the project is closely controlled by and figured in terms of one person, the Finnish programmer Linus Torvalds. Neither the minor celebrity of Torvalds, nor the relatively conventional architecture or technical features of Linux itself constitute radical innovations. So how does something such as Linux become the object of intense feeling? How does it manage to enlist hackers and programmers to work inordinately hard on it, and at the same time become something that Wall Street, West Coast venture capital, the Pentagon, the European Union, schools in Goa, the Japanese government, anti-globalization protesters in Genoa and many other institutions (but not the US Congress [NewsForge, 2002]), organizations, groups and individuals see as desirable, as a solution to their problems?

This article argues that the ongoing development of Linux can be understood as a partial solution to a more general problem concerning the relation between information infrastructures and conventions, proprietary objects, and the cultural life of code in circulation. Linux represents a form of collective agency in the process of constituting itself. This ongoing constitution is performative with respect to the efficacy of Linux as a technical object and with respect to the fabrication of Linux as a cultural entity. By virtue of the operationality of code, Linux functions as an *indexical icon*, something that recursively refers to a description of itself. Put more concisely, and this is one of the key mechanisms addressed in the article, it is 'a self-reflexive use of reference that in creating a representation of an ongoing act, also enacts it' (Lee and LiPuma, 2002: 195). The performative constitution of collective agency associated with Linux is complex. Just as the Linux kernel works by co-ordinating and scheduling computing processes of many different kinds, so too the ongoing development of Linux involves coordination and scheduling of many different programmers' work. Together the social organization of code work and the operating system itself constitute a process in which describing and enacting what is described coalesce. Importantly, performativity works by covering over and holding something in place.

The Problem of Operationality

Information networks take many different forms and rely on a variety of different channels to propagate and circulate messages (ethernet, cable, modem, wireless, satellite, radio, microwave, optical fibre, etc.). These different channels generally connect into a relatively small range of comput-ing hardware (Intel, Sparc, PowerPC) and operating system platforms (Unix,

Windows, Linux, MacOS). As Scott Lash says, platforms constitute zones on which 'technological forms of life' depend (2002: 24).[5] Linux constitutes one such platform.[6] It is an operating system currently operating at tens of millions of different points in the information networks. It constitutes a domain where problems of property, commodity specificity, communication and production are played out fairly intensely. Like other operating systems such as those produced by Microsoft or Apple, Linux puts layers, many layers in fact, of code between the loosely assembled commodity hardware (CPUs, motherboards, disk drives, network interfaces, various parts, graphics and sounds cards, etc.) and the application software (various programs and applications such as webservers, databases, word processing programs) on which user-interfaces are focused. Linux differs from operating systems sold by Microsoft and Apple in that it runs on a range of different hardware platforms, ranging from handheld computers (Shah, 2002), through game consoles such as Sega Dreamcast and Sony PlayStation to IBM supercomputer clusters. Lately Linux 'ports' (or adaptations) have appeared for various embedded or realtime platforms used in controlled non-computing devices or systems.

Apart from the incessant struggles over market share between Microsoft and Apple, and the endless debates and campaigns that struggle over the relative merits of Mac vs. Windows, the question of what operating system runs on a given hardware platform would hardly appear to be a pressing cultural question. Linux emerges from the relatively affluent domains of university computer science departments, corporate IT departments and research labs, and various software US computer companies. Linux has, however, to a certain extent, gone beyond these privileged domains of technical production. One question is how that has happened. Lash suggests that '[i]n the representational culture the subject is in a different world than things. In the technological culture the subject is in the world with things' (2002: 156). Within technological cultures, operating systems and server software constitute just such things which we might term 'culture-objects' by virtue of the density of the mediations and relationality that run through them and texture them. As culture becomes 'operational', or as information technologies become more cultural, that is, as they merge into wider circulatory practices of ordering and coding, of representing and regulating differences in some ways and not others, erstwhile infrastructural things like operating systems, protocols, algorithms and code figure as singularities.

In what sense does an operating system (or for that matter, a genome, a public database, a stem cell, etc.) constitute a culture-object? An operating system is hardly a medium, in the sense that television or newspapers are media. Nor is an operating system a message whose content or meaning can be analysed. Certainly, operating systems are often products that are packaged, circulated, regulated (as in the US Attorney General's anti-trust action against Microsoft Corporation's Windows operating system) and consumed on a massive scale. Yet at the same time, as Linux indicates, an operating system may not be reducible to a conventional commodified object if it constantly modulates as it moves through a distributed collective of programmers and system administrators. By contrast, the same thing cannot be said for computer hardware. Almost without exception, computer hardware is commodified and its production is industrial.

An operating system such as Linux challenges the typical analytical separation between production, circulation and reception (consumption, use, spectating or audiencing) in a number of ways. If much social and cultural theory relies on that separation *de facto*, the question remains: what kind of culture-object is Linux? Through what modalities (technological, compositional, social) and at what sites (production, reception, as an object in its own right) should it be analysed?

Software Performance and Performativity

As a technical object, the principal claims made for Linux have been remarkably consistent. The claims are that Linux is 'free' and that it performs better than similar commercial products such as those made by Microsoft or Sun. Leaving aside for a moment the question of whether or how Linux is 'free', Linux has been and continues to be clearly figured in terms of its technical performance. For instance, a recent advertising campaign for the database and 'enterprise infrastructure' company Oracle Corporation claims that Linux and Oracle together are 'unbreakable' (Oracle Corporation, 2002). Computer companies ranging from IBM, Compaq, Hewlett Packard, to Dell, as well as retailers such as Wal-Mart all currently sell Linux on this basis. For the most part, they market Linux to corporate and government clients. However, in some places, consumer sales of Linux have become significant. In China, where legal controls over intellectual property are beginning to be enforced as a condition of China's entry into the World Trade Organization, a distribution Linux called TurboLinux outsells Microsoft Windows (TurboLinux, 2002). We could say then that as a culture-object, Linux figures largely in terms of performance combined with low cost. How can that performance, denoted by terms such as 'unbreakable', 'proven performer' (Redhat, 2002), or 'stability', be analysed without falling back on a naturalized or technologistic understanding of performance?

Broadly speaking, the technical performance of Linux is represented within a discursive formation associated with information and communication processes. That formation discursively situates and complicates any unmediated technical 'performance' of Linux. Crucially, I am proposing that technical *performance* is coupled, in ways that need to be analysed, to *performativity*. As power becomes 'performative' ('power itself is no longer primarily pedagogical or narrative but instead itself performative' [Lash, 2002: 25]), information and communication systems and networks come to be one important venue in which power is enacted. But what does it mean to say that an object becomes performative? Given the many ways, sites and levels at which performativity works, and the large-scale differences of class, race, gender and sexuality for which the conceptual applications of performativity have been developed, how is something quite thing-like or even infrastructural like Linux at all relevant to the contemporary performativity of power?

One broad motivation driving for performativity has been the need to account for the extra-linguistic effects of linguistic praxis. Sometimes, speech moves things. Across the wide ranging discussions in social, political and cultural theory of performativity in the past decades (Derrida, 1982; Butler, 1997), this boosting effect has been continually emphasized. The idea of performativity is that utterances are always 'redoubled' (Butler, 1997: 11) by an act which cannot itself be fully recognized or made visible in the utterance. No statement is utterly dissociated from body, place or time. Utterances sometimes have a 'divine' effect, that is, they sometimes make things happen. Such occurrences attest to something indissociably adjacent to the utterance – the practices and singularities of an authorizing context which transports the speech act. Computer code, an exemplar of formal clarity and univocity, seems to be an unlikely candidate for performative analysis. Yet computer code never actually exists or operates apart from a prior set of practices which allows it to do things. Analysing Linux's performativity would involve extending a speech-based notion of code as 'instruction' to include the mediated practices of coding or programming, distributing, configuring and running an operating system. Viewed from this angle, Linux would have to be understood not just in terms of the meanings ascribed to it, or in terms of its effects on the movements of data and information in communication networks. Rather, it would be an objectification of a linguistic praxis, as a 'self-reflexive use of reference that enacts the act that it represents'. As Lee and LiPuma write:

> The analytical problem is how to extend what has been a speech act-based notion of performativity to other discursively mediated practices, including ritual, economic practices, and even reading. What is interesting about performatives is that they go beyond reference and description – indeed, they seem to create the very speech act they refer to. (2002: 193)

What would be the Linux 'act'? The insistent claims about superior technical performance and being 'free' mentioned above constitute, I would suggest, a public face of the 'speech act' in question here. Insofar as being free and being technically better matter, Linux succeeds as a speech act. However, the performativity of Linux is complex and provisional. Arguably it is not constituted principally through meaning-making processes such as narrative, but rather through what Lee and LiPuma term 'circulation'. They write:

> Performativity has been considered a quintessentially cultural phenomenon that is tied to the creation of meaning, whereas circulation and exchange have been seen as processes that transmit meanings, rather than as constitutive acts in themselves. Overcoming this bifurcation will involve rethinking

circulation as a cultural phenomenon, as what we call cultures of circulation. (2002: 192)

Lee and LiPuma argue that circulation produces performative effects. That is, processes of circulation themselves objectify linguistic praxis. They enact something. If we accept that information and communication constitute a central venue for the performativity of some important contemporary forms of power, then the circulation and exchange of software and code involved in the infrastructure of communication could well be analysed there in performative terms. From this perspective, the explicit claims about Linux's technical performance, as they appear in advertisements, editorials, newsgroups, how-to manuals and popular press accounts, would be only a secondary effect of the more primary, collective performativity of practices channelled through computer code.

Circulating, Distributing, and Co-ordinating Code

The central claim that I am advancing here concerns the set of practices that produce, circulate and consume Linux. As an operational object serving as a platform, Linux quite literally co-ordinates the circulation of specific social actions pertaining to information and communication networks. At the same time, co-ordinated actions centred on Linux constantly modulate it as an object in self-referential ways. For instance, development work done by programmers on the Linux kernel modifies the very platform on which they do their programming. The code they work on can be seen as an intricate description of the many features of a highly configurable technological ensemble. That features of that description constantly change through 'kernel hacking'. Such changes in turn affect how the ensemble itself operates. Because it operates differently, Linux as an operational act changes. Each *release* of the kernel (there have been dozens of releases) circulates differently because the features have changed. New hardware configurations, new communication protocols and new kinds of connectivity are constantly incorporated into the kernel. Through that modulatory circulation an objectification of linguistic praxis occurs which gives rise to both the effects of technical performance and the freeing up of the proprietary status of the code.[7]

The agencing effect of performatives rests on their capacity to 'objectify' some kind of praxis, typically linguistic practices. This objectification, however, only succeeds provisionally or partially. Judith Butler writes:

> If a performative provisionally succeeds (and I will suggest that 'success' is always and only provisional), then it is not because an intention successfully governs the action of speech, but only because that action echoes prior actions, and *accumulates the force of authority through the repetition or citation of a prior and authoritative set of practices*. It is not simply that the speech act takes place within a practice, but that the act is itself a ritualized

> practice. What this means, then, is that a performative 'works' to the extent
> that it *draws on and covers over* the constitutive conventions by which it is
> mobilized. (1997: 51)

Could the analysis of performatives extend to the repetition or citation of a
prior authorizing set of practices in code? How would the 'repetition or
citation' involved in Linux help us understand its success? What is cited
or repeated? What is covered over? What extra purchase would the idea of
the partial workability of performatives give us in understanding collective
investment in Linux?

'Distros': Repeating and Citing Linux

One curious feature of Linux is its propensity to circulate in many different
forms. At one end of the contemporary spectrum, we could point to recent
distributions of Linux as an artwork. An online and radio broadcast performance of Linux called RadioFreeLinux ran during the first half of 2002
(Radioqualia, 2002). The work consisted solely of the source code (the
program as written and read by programmers) of the Linux kernel (the
central part of the operating system that interfaces between the endless variations in hardware and the application programs and services that users run
on their computers) being read out line by line over various radio and
streaming web-radio sites.

   Although not particularly fascinating to listen to, the work signals that
the value of computer code ('source code') is changing. On the one hand, it
is 'free' in the sense of being freely available to be copied and distributed.
It is often given away on the CD-ROMs stuck to the front of popular
computer magazines. On the other hand, source code possesses to a greater
or less degree some social or cultural value apart from economic value. The
title *RadioFreeLinux* suggests that source code should circulate or be broadcast like news or entertainment. The work is not an isolated aberration. In
1999, Linus Torvalds was awarded a significant new media art prize at Ars
Electronica. Linux had almost become, perhaps not quite, an object of
aesthetic value:

> The Jury of the.net category awards the 1999 Golden Nica to Linus Torvalds
> as representing all of those, who have worked on this project [Linux] in past
> years and will be participating in it in the future. . . . It is also intended to
> spark a discussion about whether a source code itself can be an artwork. (Ars
> Electronica, 1999)

Even if art prizes are partly driven by desire for frequent revolutions and
innovations, what does it signify when source code becomes an artwork?
Given that source code for programs has been around for at least 50 years,
why has source code become something that people want to 'read' now?

   At the other end of the spectrum, early in 2002 Sony Corporation
announced that it was releasing a version of Linux for the games console,

PlayStation2. Typically gaming development environments such as *ProDG* for the PlayStation cost between $US5,000–10,000 for restricted developer licences (SN Systems, 2002). Sony itself licences games development in order to control the quality of the games commercially released for the console. In a shift of licensing policy, Sony announced Linux for PlayStation by saying:

> The PlayStation 2-specific libraries will be released under the LGPL; there are no proprietary licenses involved. Sony's distribution of Linux is based on Kondara, which in turn is based on Redhat. The documentation with this kit will give all the same information about the PS2 hardware that Sony provides its licensed game developers (but it won't give access to the system's anti-piracy mechanisms). This will include full details on the PS2's proprietary Emotion Engine core instruction set, the Graphic Synthesizer, and the Vector Processing Units. (Wen, 2002)

The migration of open source code onto the proprietary hardware PS/2 platform is not unique. As mentioned above, Linux has been 'ported' to many different proprietary hardware platforms ranging from PDAs to main-frames.[8] This announcement points to something else: 'Sony's *distribution* of Linux is based on Kondara, which in turn is based on Redhat'. This avers to the fact that Linux exists in many different 'distros'. Kondara and Redhat are existing distributions or repackagings of the Linux kernel along with associated software. It is also possible to build the operating system by downloading the kernel source code (www.kernel.org) and some other pieces of software independently of any of the branded distributions. While it is possible to visit various websites and ftp servers, download source code and compile it to build a complete operating systems, almost everyone using the Linux kernel relies on a distribution such as Redhat, Mandrake, Debian, SuSe, Gentoo, and so forth. Most people work with Linux by downloading a pre-packaged branded distribution such as Redhat, Mandrake or Debian, and then configuring the system to their own needs. Distributions empha-size different features. Often distributions re-circulate an existing distri-bution in a somewhat modified form. Kondara modifies the Redhat distribution in ways that adapt it to the hardware specificities of the PS/2. Thus the PS/2 distro figures as just one among around at least 100 different commercial and non-commercial distributions of Linux (Distro, 2002). Some of these, such as PS/2 Linux, have the specific purpose of making a hardware platform more widely programmable and less of a blackbox. Some concentrate on particular application domains such as 'desktop users'. The European Commission-funded *Agnula* distribution focuses on audiovisual computing. Others have specifically a national or language focus. TurboLinux for instance, as the first Linux distribution to provide Chinese language support, focuses on selling Linux in China (Turbolinux, 2002). For the past five years, new distributions of Linux have appeared and disappeared at a high rate.

How can we understand the existence of this multiplication of different incarnations of the same thing, the Linux kernel, ranging from an audio broadcast, to quasi-proprietary Sony distribution, then to dozens of branded distributions, and finally to the source code itself, readily available from many different websites, ftp servers and mirror websites? Linux circulates as artwork, as commercially packaged commodity sold by many different companies, and as freely available source code files, constantly worked on and exchanged using sophisticated software mechanisms of co-ordination, scheduling and organization running on Linux servers. The commercial distributions have largely come into existence as a way of capturing the 'free labour' embodied in Linux. Without violating the fairly stringent GPL licences which prevent the software itself being sold, companies like Redhat and IBM effectively give away Linux distributions, but charge for the support services needed to configure and maintain them in operation (Taylor, 1999). At the same time, the free work on the various components of Linux continues, and the changes constantly made to it by numerous programmers are successively incorporated into new releases of the commercial and non-commercial distributions. The economics of this process were subject to strong stock market speculation during the late 1990s (Redhat, 1999; Taylor, 1999), and were extensively analysed by academic researchers (Tuomi, 2000).

The Internet-based coordination mechanisms that permit collective development work on the kernel have been fairly thoroughly described in both semi-popular and academic accounts of open source (Bezroukov, 1999; Moody, 2001). For the moment, the important point is that the circulation of Linux in dozens of different distributions connects technical performance and the performativity of Linux as a culture-object. The distributions configure Linux so that it can circulate across different hardware platforms, and between different cultural, institutional and national domains. Each new distribution, and each successive release of an existing distribution incorporates and repeats the conventions embodied in the Linux kernel but adapts those conventions to a slightly different situation – a different hardware platform, a different language grouping, a different kind of computing task (supercomputing, office productivity, eCommerce, genomic sequence searching and alignment).

Linux: An Object out of Control?

Accounts of science and technology over the past decade such as Latour (1996), Haraway (1997) and Lash (2002) have sought to analyse why certain contemporary objects like Linux are 'out of control'. By 'out of control', they mean that these objects generate unexpected consequences. They proliferate at a rate that cannot be accounted for by theories of innovation based on diffusion. Such accounts argue that technical objects like Linux become unstable and proliferate as a consequence of the translation or objectification of social and cultural relations into and through them. As Lash puts it, objects get out of control as an unintended consequence of modern rule-regulated reflexivity:

> [R]eflexivity involves the reflexive monitoring of the object by the subject, in which the subject subsumes the object under rules . . . The more we monitor the object, the more the object escapes our grasp . . . This moment of contingency, is where the object, or the self, escapes the cognitive categories of the subject, is indeed aesthetic. (2002: 50)

As an object, the Linux kernel incorporates a tangled web of rules, conventions, standards, and protocols pertaining to information and communication networks, as well as specific features relating to the commodity computing hardware. The conventions range from the fundamental binary abstraction on which all code operations rest through to the highly conventional protocols for transactions and messages on the information networks. These conventions are defined by regulatory and standards organizations such as the ISO (International Standards Organization), the W3C (World Wide Web Consortium), IEEE (the Institute for Electrical and Electronic Engineering), and ANSI (the American National Standards Institute). The conventions and protocols on which nearly all computer code relies attempt to subsume information systems under rules that render those systems inspectable or reportable. So, for instance, Linux 'aims for POSIX (Portable Operating System Interfaces) compliance' (Linux Kernel, 2.4.x, 2002).

Do the conventions present in Linux explain its proliferation, its propensity to 'escape our grasp'? If, as Lash writes, 'the more we monitor the object, the more the object escapes our grasp', it could also be added that the object's escape from 'the cognitive faculties of the subject' is not an accidental loss of control, a result of risk and uncertainty. Rather, it flows from the performativity of circulation. If the notion of performativity has any traction in explaining 'cultures of circulation' such as those associated with Linux, it might be that it provides a different way of apprehending the proliferation and complexification of objects. My analysis of Linux as a collective formation in the process of performatively constituting itself hinges on this point. The kernel institutes a performative to the extent that in describing a nexus between commodity hardware and conventional, rule-governed orderings and movements of information, it also enacts that nexus. At core, the line between code object – the kernel – and code subject – the hackers and programmers personified in the figure of Linus Torvalds – wavers uncertainly. To be a hacker working on Linux means in some way to see Linux as a more than just an object to be apprehended cognitively, but as a form of life which involves challenging norms of property ownership and corporate organization of work (Himanen, 2001; Moody, 2001).

What Linux Covers Over

Performativity undermines the prerogative of either a subject or an object to give meaning to things. The agential effect of performativity arises first of all, as we have seen, through repetition and citation. Linux repeats itself across platforms, and in different contexts. But another dimension of its performativity needs to be analysed. Performatives also 'succeed' not only by citing, or enacting through describing, but by 'covering over', as Butler suggests, the 'authoritative set of practices' which lend force to the enacting. What 'authoritative set of practices' or 'constitutive conventions' does Linux cover over? The 'authorizing context' is the set of conventions and practices whose repetition gives a performative act its force.

Judging by what is said in advertisements, in newsgroups or in editorials and essays about Linux, there would be some justification in thinking that the history of Linux, its biography as a technical object and commodity, is brief, linear, uncomplicated and dates from around November 1991. Around this time, the Internet first begins to enter into public awareness. The history of Linux is often presented as a consequence of the coordination and circulation of code that the Internet permitted in the late 1980s and early 1990s (Moody, 2001). That is, because source code could be uploaded and downloaded from ftp servers through dial-in modem lines and PCs, source code circulated much more rapidly and widely than it had done during the 1970s and 1980s when only institutional and corporate computing facilities had network access. However, as mentioned above, Linux clones an earlier computer operating system, Unix. Drawing on Unix, Linux sediments a complicated relation to proprietary hardware and software. But Unix was and remains not just a piece of software, but also an associated set of coding, software design and system administrative practices, sometimes referred to as the 'Unix philosophy'. Linux's 'authorizing context', to use Butler's term, is complicated because it combines regulatory practices and coding-hacking practices. As an event, Linux is complex in ways that the popular narratives of the 'cloning of Unix' (Himanen, 2001; Moody, 2001; Lohr, 2002) occlude. The authorizing context for Linux includes gendered and classed practices which usually remain unremarked.

The kernel archive sites, such as ftp.kernel.org, show that Linux has been through dozens of releases cycles, and hundreds of minor updates. What started as a file of a few hundred kilobytes in size in 1991 has grown to 22Mbytes by mid-2002. If we wanted to ask how Linux begins to take on its own voice, or how it begins to accumulate agential force, we would need to look at both what that growth involves and covers over. The README textfile found in every release of the kernel source code since 1991 begins: 'WHAT IS LINUX? Linux is a Unix clone written from scratch by Linus Torvalds with assistance from a loosely-knit team of hackers across the Net. It aims towards POSIX compliance' (Linux Kernel 2.4.28, 2002).

First of all, although 'written from scratch', Linux heavily cites another operating system or software platform, Unix. So all the distributions of Linux currently circulating, and competing with various degrees of commercial and non-commercial success are versions of systems loosely grouped under the name 'Unix', 'a Trademark of the Bell Telephone Laboratories' (Kernighan and Ritchie, 1978: ix). Unix itself dates from summer 1969 when two computer scientists (Dennis Ritchie and Ken Thompson) at the Bell Telephone Laboratories in New Jersey developed an interactive operating system. Although there may be other elements and practices within the authorizing context from which Linux draws its performative force, it would be hard to argue against the centrality of the 'Unix philosophy'.[9] But the Unix philosophy is precisely the heterogeneous mixture of practices, names, strictures, conventions and habits characteristic of any performative nexus.

Authorizing Context (a): Interaction and Economy

With a view to understanding how an earlier operating system could function as the authorizing context for Linux, we could ask: why clone Unix rather than Windows, MacOS, or some other operating system? By the early 1990s, when the Linux kernel first appeared, Unix was widely used at universities and research institutions in the USA and Europe. Why was it so widely used? Without going into the technical lineages of operating systems, and the constant trade-offs and exchanges between academic computer science, computer manufacturers, the US military and telecommunications companies, the first academic paper published by Ritchie and Thompson reporting on Unix in 1974 (Ritchie and Thompson, 1974) emphasized the combination of the power of 'interactive use' and economy in relation to hardware:

> Perhaps the most important achievement of UNIX is to demonstrate that a powerful operating system for *interactive* use need not be expensive either in equipment or in human effort: UNIX can be run on hardware costing as little as $40,000, and less than two man-years were spent on the main system software. (1974: 365)

The 'interactive' character of UNIX was an 'expression of our desire for programming convenience' (Ritchie and Thompson, 1974: 374) rather than any predefined operational objectives for the system. In other words, the kind of work that programmers face was deeply embedded in the design of Unix in a number of different ways. Not only were the kinds of features and functionality available in Unix suited to programmers, Unix was itself coeval with a particular programming language, C, that became an industry standard and is used to write Linux code. In contrast to other commercial operating systems at the time, Ritchie and Thompson explicitly set out to design a system for 'programmers' convenience. Consequently, Unix took on a particular shape based on two major abstractions. These abstractions strongly affected Linux. The notion of everything as a *file* (Kernighan and Ritchie, 1978: 2), the notion of 'the process' (the other fundamental abstraction in Unix) and the provision of many tools and utilities for accomplishing specific program and system maintenance tasks still completely pervade Linux.

By the early 1990s, just before Linux emerged, Unix already had its own quite complicated history of exchange and value (see Salus, 1994). One of the standard undergraduate textbooks on operating systems at the time, Tanenbaum's *Operating Systems: Design and Implementation*, says: 'UNIX has been moved ("ported") to more computers than any other operating system in history, and its use is still rapidly increasing' (1987: 11). Unix migrated to many different hardware platforms both because of its specific technical attributes of Unix and the Bell AT&T licensing arrangements that meant users needed to work with each other to keep their systems running. The 'interactivity' of Unix (its ability to schedule and carry out more than one computing task at once) and its 'portability' are attributes central to Linux. From the outset, they were coupled together in the way Unix circulated. This combination of circumstances certainly forms a constitutive convention for Linux. The history of Unix lies at a complicated conjuncture between the organization of capital, state regulation of communication, and the technical work of programming.

Unix's circulation between 1969 and 1983 was strongly affected by an anti-trust case initiated in 1949 by the US Federal Department of Justice against Western Electric and AT&T, companies that jointly owned Bell Telephone Laboratories. Between the first public release of Unix in 1971 and 1983, AT&T had been bound by the terms of a court judgment made in 1956 to distribute Unix at a nominal fee to anyone who wanted it. As a telephone company, they were legally prohibited from profiting from computer software. At the same time, because they were legally prohibited from making any profit on an operating system such as Unix, AT&T provided no support and no fixes for bugs in the system. Unix was unsupported. As Salus writes:

> [T]he decision on the part of the AT&T lawyers to allow education institutions to receive Unix, but to deny support or bug fixes had an immediate effect: it forced users to share with one another. They shared ideas, information, programs, bug fixes, and hardware fixes. (1994: 65)

In 1983, a second anti-trust action (similar to the one mounted currently running against Microsoft) divested AT&T of the Bell Operating Companies and changed the commercial status of Unix. In the aftermath of the second anti-trust action, the newly formed AT&T Bell Telephone Labs, no longer a legal part of a telephone company, could license Unix for any amount it wanted to. The charge for the all-important source code rose steeply into the order of several hundred thousand dollars and it prohibited the source code from being studied on university courses (Salus, 1994: 190) where Unix had become a standard teaching tool for computer science students.

Unix is an important component of the authorizing context from which Linux emerges. Unix existed for a long time as something that the US government prohibited a telecommunications company from commercializing. AT&T Bell was compelled to give Unix away freely. At the same time, because AT&T could make no profit, it provided no support, so adopters of Unix were motivated to do their own maintenance and development. Consequently, Unix became a pedagogical object. It was used to teach computer science students at universities during the 1970s and 1980s about operating systems. Because it came with its own programming language, C, Unix was also a highly mutable, reconfigurable object. Computer scientists used Unix as a testbed to explore new protocols, data structures and user interfaces. Unix grew as a part of the emerging infrastructures of networked computers emerging in the late 1970s and growing during the 1980s into the Internet. This aspect of the authorizing context centres on a particular set of communication protocols: TCP/IP. According to Abbate's *Inventing the Internet*, they were first incorporated into Unix in the late 1970s, mainly through ARPA funding (2000: 133).[10] Generally speaking, the primary component of the authorizing context underpinning the performativity of Linux comes from the peculiar set of Unix practices which intimately linked producers and users/consumers of operating systems.

Authorizing Context (b): Co-ordinating and Communicating

What does it mean to say that 'Linux is a Unix clone'? In cloning a computing environment, where is the line drawn between what is cloned and what is not? Linux, according to Torvalds, was a 'program for hackers by a hacker' (Torvalds, 1991), just as Unix was, according to Kernighan and Ritchie, a system by programmers for programmers' convenience. Given that Unix had, as Tanenbaum's text book states, already been 'ported' to more platforms than any other operating system 'in history', what was at stake in having it cloned by hackers?[11]

A second crucial and interlocking component of the authorizing context which undergirds Linux's performativity resides here. Linux follows in the wake of other 'ports' of Unix to personal computing hardware such as Minix. But the first pieces of code that could recognizably be called 'Linux' were experiments with a particular feature of the Intel 80386 chip called 'task-switching'. In a 1991 interview, Torvalds said:

> I was testing the task-switching capabilities [of the 80386], so what I did was I just made two processes and made them write to the screen and had a timer that switched tasks. One process wrote 'A', the other wrote 'B', so I saw 'AAAABBBB' and so. . . . God, I was proud over that. (Moody, 2001: 36)

Although it is a clone of existing Unix systems which had become increasingly proprietary by the early 1980s, Linux took root in an individual desire to 'test the task-switching capabilities' of the i386 family of microprocessors produced by Intel. Why was 'task-switching' of interest? One of the fundamental Unix abstractions, 'the process', supports Unix interactivity. By using the task-switching capabilities of the Intel platform (which underlay and still underlies most personal computers), Torvalds could build from the ground up a concrete implementation of the fundamental process abstraction. Unix culture then could move out of the institutional and commercial environments in which it had been developed and regulated via this adaptation to a commodity consumer hardware platform. Via the appropriation of consumer computing hardware effected by Linux, the conventional object

Unix begins to metamorphose into something more dynamic and processual. Linux, we could say, transforms the consumption of commodity computing hardware by opening a window onto the 'authorizing context' embodied in Unix. How is consumption understood here? In the case of Linux, consumption of commodity occurs through coding. Through code, the proprietary hardware specificities of different computational devices and peripherals are addressed.

The first release of the kernel in late November 1990 contains a warning about relations to the proprietary devices:

> LINUX 0.11 is a freely distributable UNIX clone . . . LINUX runs only on 386/486 AT-bus machines; porting to non-Intel architectures is likely to be difficult, as the kernel makes extensive use of 386 memory management and task primitives. (Linux Kernel 0.11, 1991)

So, in the first publicly released version of Linux, a key piece of code contained in the 'main.c' file sets the operating system time from a hardware clock:

```
/*
 * Yeah, yeah, it's ugly, but I cannot find how to do this correctly
 * and this seems to work. If anybody has more info on the real-time
 * clock I'd be interested. Most of this was trial and error, and some
 * bios-listing reading. Urghh.
 */
#define CMOS_READ(addr) ({ \
outb_p(0x80|addr,0x70); \
inb_p(0x71); \
})
```
(Linux Kernel 0.11, main.c)

As the code comments say, this is 'ugly' material to work with (especially for a researcher!). The lines of code are executed when the operating system starts up. They actually refer to some other lines that relate to hardware specificity. If we trace the line 'outb_p(0x80|addr,0x70);' to their definition in another source file, we find them defined as:

```
#define outb(value,port) \
__asm__ ("outb %%al,%%dx"::"a" (value),"d" (port))
#define inb(port) ({ \
unsigned char _v; \
__asm__ volatile ("inb %%dx,%%al":" = a" (_v):"d" (port)); \
_v; \
})
```
(Linux Kernel 0.11 /include/asm/io.h)

The code becomes even uglier and less easily readable. The increasing unreadability shows something important. These instructions read and write to specific spatial locations on the Intel x86 family of CPUs (specifically, IO [input/output] ports). Some of the very first lines of the Linux kernel are very closely tied to the deeply embedded specificities of the Intel 80386 chip. Frustrated by the opacity and obscurity of the hardware, Torvalds writes in the code comments, 'I cannot find out how to do this correctly.'

During early 1991, postings on newsgroups such as comp.os.minix, soliciting programmers to work on Linux, were framed in terms of the difficult specificities of hardware devices. This is Linus Torvalds inviting people to work with on what would later be called Linux:

> Do you pine for the nice days of minix-1.1, when men were men and wrote their own device drivers? Are you without a nice project and just dying to cut your teeth on a OS you can try to modify for your needs? Are you finding it frustrating when everything works on minix? No more all-nighters to get a nifty program working? Then this post might be just for you:-) (Torvalds, 1991)

This first announcement and call for volunteers – 'men' – focuses on the all-night pleasures of low-level system programming of 'device drivers', the parts of an operating system that speak to specific pieces of hardware rather than higher level abstractions. The mention of a time when 'men . . . wrote their own device drivers' presumes, first of all, a time when men had access to the device in question. In the 10 years since, as the source code for the kernel has grown from a few hundred kilobytes (220kB) to approximately 20MB in size, code that addresses proprietary device specificities have been added to the system. As Linux is 'ported' onto different platforms, new hardware specificities have to be addressed. At the same time, assumptions about hardware specificities built into previous versions of the code have to be rendered explicit and then moved out of shared code into specialized areas.

A final feature of the constitutive conventions at work in the performativity of Linux should be mentioned, although it has not been explicitly theorized or developed in this article. The Linux kernel is deeply tied to a gendered corporeal set of practices of programming work. The mention of 'all-nighters', of a time when 'men wrote their own device drivers', reminds us that Linux is above all a program by men for men who like to play with computing hardware. The specificity of the commodity computer hardware for which Linux hackers write device drivers correlates directly with the desire to 'modify an OS for your [own] needs'. Anthropologically speaking, these desires remain somewhat unexplained.

Conclusion

How does computer code become cultural? The background to this question comes from more general accounts of the operationality of 'the information order' (Lash, 2002: 4) or network societies (Castells, 2001). In certain ways, the Linux kernel reveals symptomatic features of the contemporary organization of material culture. In production, circulation and consumption of information, in the ways that things like Linux mediate informatic sociality, and in the ways they articulate new intersections between the different sets of practices and conventions, code objects demonstrate that operationality depends on the constitution of collective agency. While the technical performance of Linux versus other operating systems is a matter of endless debate, its capacity to enact what it represents or describes, is harder to contest. The proliferation of distros, sites, enterprises, products and designs associated with Linux attests to that performative force.

The main thread of my analysis of Linux locates operationality within a specific culture of circulation of code. Extending performativity from spoken or written texts to objects is not new. Latour, for instance, implicitly relies on a concept of performativity when he writes:

> [P]rograms are written, chips are engraved like etchings or photographed like plans. Yet they do what they say? Yes, of course, for all of them – texts and things – act. They are programs of action whose scriptor may delegate their realization to electrons, or signs, or habits, or neurons. (1996: 223)

In claiming that 'programs' do what they say, Latour attributes performativity to objects insofar as they 'realize' specific social actions. It needs emphasizing that realizing programs of action is complicated and contested.

Can any thing, any culture-object, be performative or is contestation, risk, failure, partiality? The concept of performativity explains an objectification of social praxis. Yet a subjectification occurs as a correlate of this objectification. The mode of address associated with performativity is something that would merit further analysis in relation to code objects. Linux could be seen as standing at the point of convergence of practices coming from academic-industrial computer science (embodied in Unix) and the large-scale production, circulation and consumption of computers as consumer electronics. Augmenting the claim that objects act and speak by virtue of delegation, the performativity of a culture-object such as Linux might be further glossed as a *partial* solution to a problem of social identity or authority for information elites. Caught between highly commodified hardware production, the stringent regulation of intellectual property and an ethos of free-wheeling creativity and autonomous work, sometimes it seems that hackers have stabilized some things (gender norms, for instance) and radically reorganized others – the production and consumption of software. Linux constitutes a partial solution to the problem of how different authorizing contexts coming from the quasi-academic ethos of Unix computing and its coding conventions, proprietary licensing of software, the plethora of commodity computing hardware and protocols for networking can be articulated together. Linux remains a partial solution since the programming collective has been unable to overcome its own limitations in regard to gender. The most egregious feature of Linux – the fact that the credits listing is almost with exception male – shows that the objectification of social practice synthesizes some things with great novelty but keeps others static.

As a postscript to the opening scenario from Bruce Sterling's *Distraction*, as mentioned earlier, a condition of China's entry into the World Trade Organization is a strengthened regulation of intellectual property law. The growth of Linux in China, and its recent adoption by the Beijing city government, suggest that Sterling's fiction might miss the mark on this point. However, the broader point of the scenario from *Distraction* probably does hold: a transformation in the organization of material culture is involved in the emergence and dynamism of things like Linux. If 'a shift from a register of meaning to one of operationality' (Lash, 2002: 216) is underway, operationality will need to be understood not in the mathematical-technical terms that predominate in some accounts, but as an articulation of diverse realities.

Notes

1. Under the current circumstances, it becomes increasingly difficult to locate software or code objects as a part of monolithic globalizing juggernaut, flattening whole domains as it rolls across the face of the globe. Nation-states such as the USA attempt to break the monopolies of major software companies (e.g. the Department of Justice's anti-trust case against Microsoft), while self-organized groups of software developers clone whole operating systems in their 'free' time (for instance, the Linux operating system).

2. Even the term 'code' itself is subject to a multiplicity of interpretations. The *Oxford Dictionary of Computing* offers no less than 113 technical terms that use the word 'code' in the domain of computing science and information technology. In the context of this discussion, 'source code' and 'binary code' are probably the two most relevant terms among the code plethora. Programmers read and write 'source code'; the software itself – for instance, the Linux kernel – is principally 'binary code' produced by compiling the 'source code'.

3. The notion of 'technoscape' outlined by Arjun Appadurai in the framework of his influential notion of the 'ethnoscape' offers one possibility here: 'by technoscape, I mean the global configuration, also ever fluid, of technology and the fact that technology, both high and low, both mechanical and informational, now moves at high speeds across various kinds of previously impervious boundaries' (Appadurai, 1996: 34). If the well-known notion ethnoscape couples migratory movements of people to imagined movements ('going home', 'making a new life', etc.), the technoscape couples, by analogy, movements of technology with imaginings of that movement. The notion of technoscape remains relatively undeveloped by Appadurai (1996). It refers only to a 'fluid global configuration' and the movement of technology across boundaries (between countries, between workplace and home, etc.). The concept of technoscape needs to offer ways of talking about how technologies move about together with imagining how technologies move. To match the conceptual sophistication of Appadurai's understanding of the ethnoscape, the technoscape needs to countenance a topographically and materially diverse domain, marked by boundaries, thresholds and paths, populated by different imaginings, devices, infrastructures and systems in interaction.

4. The distinction between 'free software' and 'Open Source Software' has emerged

in the past five years. Free software, as promoted by the Free Software Foundation (http://www.gnu.org/) is supported by a liberal, civil rights discourse of free speech. 'Open Source Software', with much closer links to corporate and industry groups, is more pragmatically oriented approach to software development based on access to source code and the right to distribute and circulate the source code (http://www.opensource.org/docs/definition.php).

5. Servers occupy the nodal points of information networks. Servers offer different 'services': that is, they configure different points of access, and channel different kinds of movement of data through information networks. Some servers send web pages, some receive and send email messages, some transport or 'stream' audio or video. Dozens of different protocols (ip, tcp, icmp, ftp, irc, telnet, ssh, ssl, rmi, rpc, smtp, nmtp, . . .) regulate these movements of data.

6. A web page associated with this article shows a table that classifies some typical materials that circulate around and through Linux. See: http://www.lancs.ac.uk/ staff/mackenza/research/opensource_materials.html

7. The feedback loop between performance and performativity mentioned earlier plays an important role because it triggers constant modulations both in the object itself and the practices associated with that object. Linux gains increasing traction as a commodity partly on the basis of this performance, yet that performance relies on the performativity of the practices associated with Linux.

8. Sony is giving developers access to the 'proprietary Emotion Engine'. On the one hand, games platforms are centred on the production and manipulation of affect through visual, audio and tactile events. In game play, certain kinds of labour 'in the bodily mode' heavily interact with audiovisual (and tactile) information objects. Some of the most dense and complicated interfaces to information systems are found in gaming. On the other hand, as the fine detail on the PlayStation Linux site shows, Sony is not giving any programmatic access to the DVD or CD drives. Understandably, they are trying to prevent the PlayStation from becoming a DVD ripper, something that would undermine Sony's broader audiovisual entertainment business.

9. At the time of writing this article, a Unix company, SCO, has threatened massive legal action against *all* commercial uses of Linux. SCO's argument is that Linux contains hidden Unix code and therefore constitutes an infringement of intellectual property rights vested in the current owners of the original Unix licences, SCO.

10. During the past two decades, these protocols have come to almost universally regulate the flow of data between hosts or servers on the Internet. These protocols provided a way of connecting many computers together, and ordering the movement of data (packets) between different parts of the network without making many assumptions about the content or internal structure of the data.

11. Expressive traits of Unix enter Linux partly through university undergraduate courses on operating systems. Linux bears the marks of Linus Torvalds' close reading of Tanenbaum's academic textbook on operating systems during 1990. In the mid-1980s, Tanenbaum, a computer science academic in Amsterdam, wrote a Unix clone called Minix as a companion to his textbook and in order to be able to teach students about Unix as an operating system without incurring AT&T's newly imposed licensing fees. The source code of AT&T's Unix was not available to students. By contrast, the source code for Minix was publicly available for a small fee so that students could understand how the system worked. Thus, by 1990, as an undergraduate computer science student at Helsinki University, Linus Torvalds himself had access to a pre-existing clone that could run on an Intel-based PC, and

a textbook that gave a complete account of the architecture of a Unix operating system to explain it. But why then did Linux come into existence, given that Minix already ran on personal computers and provided Unix-style computing to almost anyone, students included, for a small fee? Tanenbaum's Minix (which had tens of thousands of users) was a system designed to support teaching about operating system design. It downplayed hardware specificities so that the system would stay simple and easy to understand. It was only slowly enhanced to deal with a wider range of hardware devices. If Minix emerged in response to the licensing restrictions on Unix source code, Linux was defined in relation to the proprietary nature of hardware. Linux was a system that transformed a didactic project with restricted circulation (Minix) into a widely circulating and rapidly modulating object.

References

Abbate, J. (2000) *Inventing the Internet*. Cambridge, MA: MIT Press.

Appadurai, A. (1996) *Modernity at Large: Cultural Dimensions of Globalization*. Minneapolis: University of Minnesota Press.

Ars Electronica (1999) 'Golden Nica for Linux/Linus Torvalds (Finland) "LINUX" ', http://prixars.aec.at/history/jurybegr/1999/E99www.html (accessed 12 November 2002).

Benjamin L. and E. LiPuma (2002) 'Cultures of Circulation: The Imaginations of Modernity', *Public Culture* 14(1): 191–213.

Bezroukov, N. (1999) 'Open Source Software Development as a Special Type of Academic Research', *First Monday* 4(10), http://www.firstmonday.org/issues/issue4_10/bezroukov/index.html (accessed 14 November 2002).

Bowker, G.C. and S.L. Star (1999) *Sorting Things Out: Classification and its Consequences*. Cambridge, MA: MIT Press.

Butler, J. (1997) *Excitable Speech: A Politics of the Performative*. London: Routledge.

Castells, M. (2001) *The Internet Galaxy: Reflections on the Internet, Business, and Society*. Oxford: Oxford University Press.

Derrida, J. (1982) *Margins of Philosophy*. Brighton: Harvester Press.

Distro (2002) http://www.distro.org (accessed 12 November 2002).

Haraway, Donna J. (1997) *Modest_Witness@Second_Millennium. FemaleMan©_Meets_OncoMouse™: Feminism and Technoscience*. London: Routledge.

Himanen, P. (2001) *The Hacker Ethic and the Spirit of the Information Age*. London: Secker & Warburg.

Kernighan, B. and D. Ritchie (1978) *The C Programming Language*. Englewood Cliffs, NJ: Prentice Hall.

Lash, S. (2002) *Critique of Information*. London: Sage.

Latour, B. (1996) *Aramis, or the Love of Technology*. Trans. C. Porter. Cambridge, MA: Harvard University Press.

Lee, B. and E. LiPuma (2002) 'Cultures of Circulation: The Imaginations of Modernity', *Public Culture* 14(1): 191–213.

Linux Kernel 0.11 (1991) "/include/asm/io.h" ftp.kernel.org

Linux Kernel 2.4.28 (2002) ftp.kernel.org

Linux Kernel 2.4.x (2002) 'README.txt' http://www.kernel.org (accessed 22 January 2004).

Lohr, S. (2002) *GOTO Software: Software Superheroes from Fortran to the Internet Age*. New York: Profile Books.

Moody, G. (2001) *Rebel Code: Linux and the Open Source Revolution*. Harmondsworth: Penguin.

NewsForge (2002) 'Washington State Congressman Attempts to Outlaw Open Source', http://newsvac.newsforge.com/newsvac/02/10/23/1247236.shtml?tid=4 (accessed 2 December 2002).

Noronha, F. (2002) 'Open-Source Software Opens New Windows to Third-World', *Linux Journal*, http://www.linuxjournal.com/article.php?sid=6049 (accessed 3 May 2002).

Oracle Corporation (2002) 'Unbreakable Linux', *The Economist*, 21–7 September.

Oxford Dictionary of Computing (1996) Oxford: Oxford University Press. *Oxford Reference Online*, http://www.oxfordreference.com/views/ENTRY.html?subview=Main&entry=t11.e4940 (accessed 20 January 2004).

Poster, M. (2001) *The Information Subject: Essays*. Amsterdam: G & B Arts.

Radioqualia (2002) http://radioqualia.va.com.au/freeradiolinux (accessed 2 December 2002).

Redhat (1999) 'Venture Communism', *The Economist*, 12 June.

Redhat (2002) http://www.redhat.com (accessed 24 September 2002).

Ritchie, D.M. and K. Thompson (1974) 'The UNIX Time-Sharing System', *Communications of the ACT* 17(7): 365–75.

Salus, P. (1994) *A Quarter Century of UNIX*. Reading, MA: Addison-Wesley.

Shah, R. (2002) 'The Penguinista's PDA', *Linux World*, http://www.linuxworld.com/linuxworld/lw-2000-06/lw-06-h3600.html

SN Systems (2002) http://www.snsys.com/PlayStation2/ProDG.htm (accessed 2 December 2002).

Sterling, B. (1998) *Distraction*. London: Victor Gollancz.

Tanenbaum, A.S. (1987) *Operating Systems: Design and Implementation*. Englewood Cliffs, NJ: Prentice Hall.

Taylor, R. (1999) 'Red Hat Shares Surge on Debut', *The Financial Times*, 'Companies and Finance: International', 12 August: 20.

Torvalds, L. (1991) *Subject: Free Minix-Like Kernel Sources for 386-AT*, 5 October 1991, 08:53:28 PST 1991 (accessed 24 July 2003). Available from Newsgroup: comp.os.minix

Tuomi, I. (2000) 'Internet, Innovation and Open Source: Actors in the Network', *First Monday*, http://www.firstmonday.org/issues/issue6_1/tuomi/index.html#author (accessed 12 November 2002).

TurboLinux (2002) 'Turbolinux Outsells Microsoft Windows in China', http://www.turbolinux.com/news/pr/federal.html (accessed 25 November 2002).

Wen, H. (2002) 'Opening up the PlayStation 2 with Linux', http://linux.oreillynet.com/pub/a/linux/2002/03/21/linuxps2.html (accessed 21 March 2002).

Adrian Mackenzie is currently studying the cultural trajectories of code. He has previously worked on theories of technology, embodiment and temporality (*Transductions: Bodies and Machines at Speed*, Continuum [2002]).

'Contemplating a Self-portrait as a Pharmacist'

A Trade Mark Style of Doing Art and Science

Celia Lury

'Becoming a brand name is an important part of life', says Mr Hirst. 'It's the world we live in.' (*The Economist*, 10 February 2001)

Introduction

IN THIS statement, the artist Damien Hirst proclaims the importance of being a brand name, an identity that has also been imputed to young British artists (yBas or NBA[1]), whose acronymic titles are sometimes taken as an aspiration to global corporatism:

> Britain is developing macro-brands – whole industries where the word 'British' raises the value of the product. British film, British fashion, British art and British architecture are more fashionable than ever. Damien Hirst is a thriving brand whose name adds immense value to a product.
>
> What's hot
>
> Vodafone, Manchester United, Virgin, Conran, WPP, Egg, Richard Rogers, Brit Art, Hussein Chalayan, Sky, First Direct, Orange, Royal Bank of Scotland
>
> What's not
>
> Rover, Marks & Spencer, the BBC, British Airways, Laura Ashley, Mulberry, Liberty, British Telecom, Barclays. (Arlidge, 2000)

Historically, the brand name was a mark of ownership intended to create trust in the consumer as a guarantor of the quality of particular products. And trade mark law was the means to secure a monopoly on the use of that mark with the dual purpose of protecting the owner from unfair competition and the consumer from 'confusion' as to the origin of the good. But increasingly the brand name is not the mark of an originary relationship between producer and products but is rather the mark of the organization of a set of relations between products in time (Lury, 2004). In the shift away from the attribution of value in relation to an origin, the contemporary brand is a mark of a transformation in the author function. This is the principle that both Michel Foucault ('What is an author?', [1969] 1998) and Isabelle Stengers ('Who is an author?', 1997) propose as one of the most important ways in which we discriminate between texts or fictions in art and science. Foucault writes:

> [The author function is] the result of a complex operation that constructs a certain being of reason that we call 'author'.[2] Critics doubtless try to give this being of reason a realistic status. . . . Nevertheless, these aspects of an individual which we designate as making him an author are only a projection, in more or less psychologizing terms, of the operations we force texts to undergo, the connections we make, the traits we establish as pertinent, the continuities we recognise, or the exclusions we practise. All these operations vary according to periods and types of discourse. We do not construct a 'philosophical author' as we do a 'poet', just as in the eighteenth century one did not construct a novelist as we do today. (1998: 213–14)

The character of the transformation in the author function that is represented by the emergence of the brand name is, as Hirst puts it, a response to 'the world we live in', that is, to a world that increasingly comes into existence as media. This is not quite the same as saying the emergence of a world that comes into existence in mediation, an activity that is seen by some as fundamental to all world-making, old and new (if not modern). As Stengers summarizes Latour: 'What comes first, then, is the activity of mediation, which not only creates the possibility of translation but also "that which" is translated, insofar as it is capable of being translated' (2000: 98).

Certainly, the brand is an organization of the activity of mediation; that is, it is an organization of an activity that 'enables us to retain the only two characteristics of action which are useful – i.e. the emergence of novelty with the impossibility of *ex nihilo* creation' (Latour, 1996: 237). But more than this, it is an organization of an activity in which the possibilities of translation, and of 'that which' is translated, are currently being transformed as the conventions that organize the specificity of particular mediums (Krauss, 1999) and mediators (Latour, 1996) are reconfigured. Another way of saying this is that the media are increasingly meta-media or meta-mediators, or perhaps, that all media are more and more mixed media.[3] What this piece seeks to address – taking Hirst as a 'figure case' (Stengers, 2000: 17) – is how the brand, as a response to the emergence of the world we live in as media, transforms 'the conditions of the emergence of novelty'. In short, the aim is to address how the author function has changed today.

The Spot Paintings

First, then, what Hirst is not. He is not, in Foucault's classic formulation of the romantic author or artist, 'the figure that, at least in appearance, is outside [the work] and antecedes it' (Foucault, [1996] 1998: 205). In the work that is marked by the name Hirst, Hirst the individual subject is not only not positioned as a prior origin, but originality itself is renounced, rendered irrelevant. For example, Hirst describes a series of works informally known as the spot paintings in the following way: 'The gaps between the spots are the same size as the spots. That's one constant factor. The spots are painted with household gloss. That's another' (Hirst, quoted in Leith, 1999). The paintings are thus not only formulaic, they are also not the product of Hirst's own labour, but are made by his assistants. Hirst tells his assistants what size he wants the paintings to be and they make them. Here then, Hirst's name may be detached not only from a subjective, interior life but also from the indexical form of the signature (Frow, 2002); it is the name as a taxonomic function alone (Gandelman, 1985). In another example of the renunciation of originality, it was recently reported that:

> The artist Damien Hirst has agreed to pay an undisclosed sum to head off legal action for breach of copyright by the designers and makers of a £14.99 toy which bears a remarkable resemblance to his celebrated 20ft bronze sculpture *Hymn*, bought by Charles Saatchi earlier this year for £1m. . . . The artist has also agreed to restrictions on future reproduction of the poly-chromatic bronze figure, described by one critic as 'a masterpiece' and 'the first key work of British art for the 21st century', which Hirst admitted was inspired by his son Connor's anatomy set. (Dyer, 2000)

Hirst had anticipated such a course of legal action; that is, he acknowledged the work was copied. In a discussion of the work in an interview, he stated, 'I might even get sued for it. I expect it. Because I copied it so directly. It's fantastic' (in Burn, 2000). In yet another example, Hirst describes organizing the show 'Some Went Mad, Some Ran Away . . .' at the Serpentine Gallery in 1994, as 'like organising already organised elements' (in Wilson, 1994: 7). More generally, he says:

> the humour [of my work] is a by-product of thinking about creating meaning through the relationships of objects. So if you're moving things round in your mind a lot, then you're looking for this fundamental, poignant relationship of two things or three things or more that says something that you're after. (In Burn and Hirst, 2001: 212)

In these examples, then, Hirst is not so much concerned with being an origin or even with originality, as with the use of a name to mark an organization of relations between things, with assemblages and reassemblages, appropriations and incorporations (Rabinow, 1998).

All this is not to say that Hirst is the first or the only artist to challenge the necessity of originality to artistic practice. Nor is it to deny the special significance of Andy Warhol (who set up his own business, Andy Warhol Enterprises, together with an industrial workshop he called Factory that (mass-)produced pictures of consumer goods such as Campbell's soup cans and boxes of Brillo pads) or that of post-modern artists such as Jeff Koons, much of whose work presents itself as a luxury commodity. Indeed, in an article that discusses the work of Jeff Koons and others (1992), Martha Buskirk argues that the author function is being subsumed by brand loyalty. Thus she suggests that the successful artist is increasingly able to establish 'his or her sole right to a particular style or method – a "trade mark" style – and others who attempt to use the same means are dismissed as mere imitators' (1992: 107). She identifies a number of factors as relevant to this development. They include the use of copyright and trade mark principles in the nexus of judgements regulating the art world; a set of unwritten understandings and agreements that extend beyond the bounds of the purely legal; and the increasing recognition accorded to artistic styles based on a conceptual working method. But while recognizing the importance of Buskirk's argument, the argument proposed here is that Hirst's practice marks a further step in the development of a trade mark style. This step is a consequence of other changes in 'the world we live in'.

Let me return to the example of the spot paintings to develop this claim. In an interview in 1999, Hirst said he wanted to stop making the spot paintings eventually, but then in 2001 he was recorded as saying 'I think I'll always make them.' Then again, he has said, 'I want them to be an endless series, but I don't want to make an endless series. I want to imply an endless series' (Hirst, 1996). In 1999, the assistants had painted about 300 (Leith, 1999); in 2000, he said that he was planning to bring all 300+ together for the first time:

> Twelve rooms. It's going to kill people dead. I think kids will run round screaming, loving it, and I think other people will get lost, like in a maze. . . .
>
>   It's an amazing fact that all objects leap beyond their own dimension. Nothing will stay still and stay in its place. Everything just jumps out and grabs you. That's how advertising works. (Burn, 2000: 220)

By 2001 about 400–500 spot paintings had been produced (*The Economist*, 2001). In viewing, each individual spot painting is simultaneously present and absent, caught up both in their own internal movement and in the movement of the (implied) endless series of which they are a part. As Hirst himself puts it, the paintings 'are full of life' (Hirst, 1996):

> I really like painting them. And I really like the machine, and I really like the movement. The movement sort of implies life. Every time they're finished, I'm desperate to do another one. The moment they stop, they start to rot and stink. (In Burn and Hirst, 2001: 221)

The spots also cover the surface of shoes (the designer Manola Blahnik's[4]), a Mini that was placed on the stairs at the entrance to the Saatchi Gallery, and a boat that ferries passengers along the Thames between Tate Modern and Tate Britain (on which the spots are of varying size but appear to be of the same size when the movement of the boat is seen from the river bank). In short, the spot paintings (and related works or products) are a series – of 400 or 500 units – characterized by their manipulation of speed, variability and miscellaneity.

This is a staging of the experience of flow as theorized by Raymond Williams (1974), an experience that is increasingly held to characterize 'the world we live in'. Williams came to formulate his influential understanding of flow in a discussion of the cultural form of television. As part of this discussion he notes the historical decline of the use of intervals between programmes in broadcasting. Or rather, he draws attention to a fundamental re-evaluation of the interval.[5] The interval, he says, no longer forms part of a sequence as the ending of one unit or the beginning of another. In the early days of broadcasting on radio, for example, there would be intervals of complete silence between programmes. But now, no longer dividing discrete programmes, no longer a silence or even an interruption, the marking of the interval – with an advertisement, a trailer, or a broadcasting company 'ident' – makes a sequence into a flow. The organization of relations between things is no longer the published sequence of discrete programmes, but a series of differently related units, some larger and some smaller than the programme, a shifting (set of) series of units, of products, images and events. In short, Williams argues that a transformed use of the interval produces an organization of time as a serial assembly or shifting series of units characterized by speed, variability and the miscellaneous as described earlier. And to the extent that the name Hirst marks the intervals between spot paintings, it too may be understood as an ident, a logo, or a brand name. It is an organization of the relations between art works (or products) in terms of the experience of flow. Hirst himself says of the spot paintings that they are

> . . . a billion times better than [the collages], in terms of getting colour to sing. It's just, like, a very exciting discovery, where you get this scientific formula that you add to this sort of mess. And also you wanted to create a logo for yourself as well. I was totally aware of what everyone else was doing. (In Burn and Hirst, 2001: 126)

Life/Live

But there is more to a brand name than the lack of an origin, the renunciation of originality, and the use of a name or sign to mark the organization of relations between works or products, at least if the name is to be protectable as a trade mark. A name must be recognizable as distinctive in law if it is not to be merely one of the many signs that mark the exploration of connections and relations between things, but a sign that carries the ability to secure ownership of a particular organization of these relations. In other words, it is in law that the ability of a brand name to be recognized as a distinctive trade mark style is really established. What might be distinctive about Hirst as a brand name – so it is suggested here – is captured in the title of one of the many exhibitions in which his work has appeared, *Life/Live*.[6] This is to say that the Hirst brand is immediacy, 'life as it is lived', living-ness. As Hirst describes the show 'Some Went Mad, Some Ran Away . . .':

> the exhibition will be about that kind of dealing with life when things are, in a way, out of control, where there is this over-layering and over-lapping of meaning giving this intense activity which is like life to me, like a writhing pit of snakes. (In Wilson, 1994: 9)

In short, the distinctiveness of the Hirst brand may be characterized as the staging of a lived relationship – 'see it, be surprised, live it and like it (or not)' (Massumi, 2000: 188).[7] This is demonstrated by the many vitrines Hirst has produced, including the celebrated cow in formaldehyde (*The Physical Impossibility of Death in the Mind of Someone Living*). But it is also exemplified in *A Thousand Years* – in which flies emerge from maggots, feed on a rotting cow's head, and die, electrocuted on a electric bar (Hirst describes the conception of this work in terms of the thought, 'What if I had a life-cycle in a box?' [in Burn, 2000: 128]). And in *In and Out of Love*, in which newly hatched butterflies also emerge only to be seen to die, this time by getting stuck in freshly applied paint. A further instance is provided by *The Last Supper*, a suite of screen prints, in an edition of 150. The prints emulate the graphic design of medicinal industry packaging.[8] In addition to the product number and dosage information, each print has the name of commonplace foods such as 'Sandwich' and 'Cornish Pasty' substituted for a drug name. The drugs selected for reproduction are mainly those used to treat cardiac trauma, AIDS and terminal conditions; they are sustenance for living in the face of death.

Defining the distinctiveness of Hirst as a brand name in these terms draws attention to the way in which Hirst's practice offers a commentary on contemporary science, especially the bio-sciences. Michael Corris (1992) notes:

> Hirst's ostentatious recapitulation of insect behaviour [in *A Thousand Years*] might also seem to burlesque the scientific method, to be a subtle indictment of the ideology of science, with its grotesque logic of induction and its equally grotesque 'controlled' animal experiments on the order of 'What would happened if we smashed a primate's skull in with a hammer?'

But could it be that Hirst's practice – especially insofar as it is understood in relation to the claim to be a brand name – is not simply a commentary on science, but is scientific practice, is science too? The name of his company, after all, is not Factory but Science. Perhaps here is an indication that Hirst believes that his work operates in 'the world we live in' in the way in which science does.

There are a number of points that might be made to support this claim, some relating to ways of doing art and others to ways of doing science. First, as a brand name, Hirst is not only positioned as the figure that precedes his work, but instead emerges in a relationship that, in his discussion of the author function, Foucault describes as characteristic of the discourse of science. Foucault outlines it thus:

> the act that founds the work [the act of creativity, discovery or invention] is on an equal footing with those that bring about its future transformations; this act becomes in some respects part of the set of modifications that it makes possible. (Foucault, [1969] 1998: 218)

Second, the elements in Hirst's work pre-exist his use of them: the things he puts together are frequently ready-mades or found objects. 'Imagine a world of spots', he says. 'Every time I do a painting a square is cut out. They regenerate. They're all connected' (Hirst, 1996). He also says, 'If you look at things in the real world under the microscope, you find that they are made up of cells. I sometimes imagine that the spot paintings are what my art looks like under the microscope' (in Burn and Hirst, 2001: 119). Third, Hirst's staging of these things in relation to each other may, in some respects, be considered experimental. Stengers, for example, describes experimentation in terms of the submission of a phenomenon to a process in which it is actively created and re-created. This process is, she says, a purification, the elimination of 'parasitic effects' that aims to make the phenomenon capable of speaking its truth. Hirst says:

> I have a built-in thing in me. I don't just solve formal problems with colour. I solve physical problems with colour. I solve all problems with colour. That's why I am more like Matisse.
> I can show you. I can show you now. Get two bits of coloured card and move them apart against white and you don't have to do anything but watch. And there's a point where you just go, '*Abso – fucking – lutely.*' And they just go there. And no one can argue with it. I win hands down. I just feel it. It's in my bones. (In Burn and Hirst, 2001: 125)

Fourth, very often in Hirst's work such manipulation or experimentation spills over into an event ('I can show you. I can show you now.'). In such cases, the value (added) of his work – understood in terms of 'the emergence of novelty' – is the exploration of the openness of connection, of interaction, 'a global excess of belonging-together enabled by but not reducible to the bare fact of having objectively come-together' (Massumi, 2000: 191).[9] As Hirst himself puts it, when describing the exhibition, 'Some Went Mad, Some Ran Away . . .': 'It's like there is a lot of energy when the aeroplane explodes and there is a lot of death after it has exploded but it is at the point when it is exploding, and I quite like it on that level' (in Wilson, 1994: 7).

Of course some commentators suggest events are increasingly, or indeed have always been, the way in which science is done (Stengers, 1997; Rabinow, 1998; Massumi, 2000). Stengers writes:

> Rather than speaking of approach, one should speak of the *event* that consti-
> tuted, and that has subsequently continued to constitute, on every occasion
> that it occurs, the practical discovery of the *possibility* of submitting a
> phenomenon to experimentation. (1997: 163, italics in original)

So perhaps Hirst does science as well as just doing art.

But Hirst does not in fact represent himself as a scientist, but more
precisely, as a pharmacist. Not only is *Contemplating a Self-Portrait as a
Pharmacist* the title of one of his best-known works, he was also one of the
owners of a restaurant called Pharmacy for a number of years. And as noted
above, he describes the process involved in producing the emergence of the
experience of colour in the spot paintings as like that of a scientific formula;
he also describes their action on viewers as like that of (anti-depressive)
pills: 'But it's like, unfailing. It's totally unfailing. The spin paintings are
shit in comparison. The spot paintings are an unfailing formula for bright-
ening up people's fucking lives' (in Burn and Hirst, 2001: 119).

Then, again, the practices of pharmacy themselves may increasingly
be understood as science conducted under the sign of the brand name. Take
this example of the transformation in the terms in which scientific innovations are made:

> A good example of competitive outflanking is made by the Reckitt & Colman
> OTC [over the counter] brand Gaviscon in the . . . gastrointestinal market. At
> first sight, its different product type . . . and mode of action (which prevents
> acid reflux in the oesophagus, rather than neutralizing acid lower down in the
> stomach) appears to offer little difference, in terms of practical end result for
> the consumer with indigestion . . .
>
> However, by developing communication around a unique problem,
> such as heartburn (essentially a layman's description of higher up stomach
> pain), Gaviscon was able to position itself as a remedy quite distinct from
> antacids. Normally, a head-to-head fight against a brand as strongly defended
> as Rennies . . . would have been fruitless for a newer entrant such as
> Gaviscon. But by choosing to fight on different territory to the antacids, by
> inventing a new problem and offering a new solution,[10] Gaviscon built a new
> OTC franchise. The key to success was its maximization of product differ-
> ence, linked to differentiation at key indication level, and exploiting this
> through consistent communication to pharmacists and consumers. (Ferrier,
> 2001: 65)

The pharmaceutical industry was (and to some extent still is) a research-
driven industry in which research was primarily conducted by medicinal
chemists in association with pharmacologists and biologists. However, the
industry is currently experiencing a complex set of related transformations
(see Barry, this issue). These include the development of genomics, combinatorial chemistry and predictive technologies that have massively
increased the number of potential compounds that are available to be evaluated in the drug discovery process. At the same time, pharmaceutical

companies are increasingly making use of computer modelling to conduct drug trials, thus enabling the creation and testing of drug compounds whose use is yet to be identified. These developments have also coincided with a greater orientation towards consumers and groups of patients, including patients' organizations, and an increasing emphasis on marketing on the part of big pharmaceutical firms, often linked to direct advertising and the promotion of so-called over-the-counter drugs. In other words, it is increasingly possible in pharmaceutical practice to identify the potential of products (or potential products) in the dis-intermediating and re-intermediating (the mixed media) practices of the brand; that is, the brand increasingly provides the experimental conditions for 'the emergence of novelty'. As one marketer puts it, 'The challenge for marketers nowadays is to introduce new products, relevant to today's needs, which will become the new brands of tomorrow' (Ferrier, 2001: 60).

What is suggested here then is that *insofar as Hirst is a brand name*, he may be said to be doing both art and science, or at least pharmacy. This is not to say that the conduct of experiments in the media(tion) of relations between things that is organized by the brand is all there is to the doing of either art or science. That is, the emergence of a new way of doing art and/or science does not define the singularity of either art or science. Instead, what is described here in relation to both art and science is the emergence of a (trade mark) style, a term used by not only Buskirk but also Stengers. The key point here, as Stengers identifies, is that the emergence of new styles may be understood as the 'proliferation of new links between science [and art] and fiction that complicate the question "Who is the author?"' (1997: 168). In the figure case of Hirst described here, the new links that complicate the question as to who or what is the author are those of the media (rather than artistic mediums or scientific mediators), and so it is to these I now return.

Black Boxes and White Cubes

In a discussion comparing the studio with the laboratory as described by Latour (1987), Svetlana Alpers (1998) argues that historically the studio is 'where the world as it gets into painting' is experienced by the artist. Or to put this another way, it is the space that enables the artist's withdrawal from the world to attend to the effect of the play of light. As such, it is not only a place to construct experiments with light ('Studio light is light constrained in various ways' [1998: 407]), but also enables the investigation of objects in relation to each other in light. This investigation is evident, for example, in the pictorial tradition of the still life: 'Objects are related to each other by reflections off neighboring things on the table – a lemon reflected onto a pewter dish, lit pewter back onto the lemon' (1998: 407). Furthermore, the studio organizes the light the artist is in; it is what enables the individual artist's experience of the world to be staged as if it – experience – is 'at its beginning' (1998: 409).[11] But the studio is not without its problems as a workplace, not the least of which is that its framing of space leaves so much of the world outside, with the artist in isolation inside.[12] Alpers describes a number of historical responses by artists to this problem, most of which involve trying to make connections, to open the studio, or to bring the outside inside the frame of the studio. One response was to see the finished works as emissaries of the artist; thus Poussin 'conceived of his paintings not as sold, but as circulating among friends'. Alternatively, Horace Vernet brought a horse into the studio, while Courbet brought a cow inside. The studio has now, however, lost its pre-eminent position: although artists still continue to work in studios (sometimes called workshops or even factories, as in the case of Warhol),[13] they also work routinely in other spaces. So, for example, Hirst first made his name by organizing three shows of his own and other artists' work in disused warehouses in London; later he was one of the artists whose work was shown in/as re-constructions of studios at the exhibition *Life/Live* in Paris. These are spaces not only of production, but also of distribution and exchange, they are the spaces of media.[14]

In a series of interviews with Hirst, interviewer Gordon Burn comments, 'The studio's like a big box over all the other boxes.' Hirst responds, 'Well, everything's a box, isn't it? Everything's a box.' For Hirst then, there are boxes within boxes; boxes – vitrines and installations – inside boxes – studios – inside boxes – galleries and museums. But what are boxes? At the most basic level, boxes divide, they separate an inside from an outside. Burn notes that, 'They [the vitrines] were originally made to keep the flies in, keep the formaldehyde in', while Hirst adds:

> And keep people out as well. You can keep them out that way, or you can keep them away with it being dangerous, like the glass pieces. I don't have any problem when it's sheer bits of glass [the skeletons]. I think they can defend themselves. But when you start getting little bits of cigarette ash and fag butts . . . When you're putting things so carefully together like that, every five minutes you're guaranteed somebody's going to be sticking a Coke can on the table. And it has a completely different meaning. Anything personal on there's going to fuck it up. (Burn, 2000: 155–6)

Sometimes the boxes are white cubes – the description given in the 1960s by Brian O'Doherty to the minimalist exhibition spaces of modernism (O'Doherty, 2000). White Cube is also the name given to the art gallery established by Jay Jopling in 1993, in which the early work of Hirst and a number of other yBas was exhibited. It is quite literally a small white cube, designed by the architect Claudio Silvestrin to create an intimate space in which a single important work or a coherent body of work could be presented within a focused environment.

White cubes are almost the obverse of the black boxes of technology of which Latour (1987) speaks, in which the object is defined by its input and output functions alone.[15] As Stengers describes it, a black box is

> an apparatus that establishes, between the data that enter it and the data that come out, a relation whose signification no scientist would want to contest,

> for to do so he would have to oppose himself to a disparate crowd of satisfied users and to rewrite entire chapters of numerous disciplines. (2000: 102)

In white cubes, it is what happens inside that is made visible, not what goes in or what comes out. But sometimes, the boxes which Hirst's work occupies are white cubes that have one or more sides which are communicative surfaces (Moor, 2003). These surfaces may be a shop-window, the cover of a magazine or book, the screen of a television, or the interface of a computer. It is almost as if the action of the communicative surface, the frame or screen is such as to turn the white cube of art into a black box of technology, and vice versa, or so the example of the OTC drug described above may be taken to suggest. Hirst comments,

> People believe science, but they don't believe art, which to me is stupid. I remember looking in a drugstore window in a bad part of London and seeing all these people looking like shit, queuing up to get their pills. Everything looked clean and perfect except the people. But there's this massive confidence in the ability of pharmaceuticals to give immortality. I thought, 'I wish people would believe in art like they believe in this.' Eventually I just said fuck it, and I put it back on the wall. It's a piece called *Pharmacy*. (Hirst, 1995)

The effect of these communicative surfaces is such that light does not shine *on* objects in the spaces in which artists and scientists work (as in the studio described by Alpers) but *through* them (McLuhan, 1996). Yet while appearing to offer transparency, they do not let the viewer see how the work inside is produced; rather, they show-case the work as its effects. They are the surfaces of a machine (a white cube, a black box) for the simulation of innovation; they act as an interface[16] (Simon, [1969] 1981; Lury, 2004). Describing *The Void*, Hirst says, 'They look like real pills; that's all that matters', and then goes on:

> You know, you've got a cow's head that looks convincing from a few feet, it's got flies all over it, it doesn't matter whether it's real or not. Because the whole dilemma is: Is it real or isn't it? It's like: Are you real? What the fuck's going on? (In Burn and Hirst, 2001: 116)

Then again, just as a shot in cinema is the mobile section of an object (Deleuze, 1986), so are the communicative surfaces described here. While, on the one hand, they may be turned inwards towards their object (life), on the other, they are turned outwards towards an expanding whole of relations that changes in time (a medium, light, life as it is lived). This, indeed, is just another way of saying that the object of Hirst's work is the world we live in as media(tion). What the claim to be a brand name by Hirst reveals, however, is that what is crucial in this turning outwards is how movement is organized as intervals in relation to the sign. Or to put it the other way round, how the sign (the logo or brand name) is differentiated by the interval, that is, how it is organized as flow, how it is organized in terms of speed, variety and the miscellaneous.

The claim to be a brand name thus calls attention to the politics of *interest* in both art and science, or what Stengers (1997, 2000) calls '*inter-esse*', that is the 'to be situated between'.[17] As Stengers puts it:

> Whom will it interest – that is, who will agree to . . . let it 'be between' (*inter-esse*) his own questions and those that produced it. This is a crucial question, because what we have to call the creation of a reality depends on it. Indeed, reality is of course not what exists independently of human beings, but that which demonstrates its existence by bringing together a multiplicity of disparate interests and practices. (1997: 165)

In French, *inter-esse* means not only *faire lien* but also *faire écran*, typically translated as 'to make a link between' and 'to stand in the way of' (Stengers, 2000: 94, although it is also worth noting that *écran* means 'screen'). This double explication of *inter-esse* provides the possibility of saying something more precise about trade mark style as a particular approach to doing art and science. What characterizes the communicative surfaces of the white cubes/black boxes described here is that they enable both these meanings of *inter-esse* simultaneously.

On the one hand, the reality that demonstrates its existence in the experimental conditions provided by the brand is one that is brought into being through an open-ended proliferation of links. In the theoretico-experimental style of science emphasized by Stengers, 'The scientist, as an author, is not addressing himself to readers, but to other authors' (1997: 161). In contrast, in the trade mark style that is described here, the politics of interest is a matter of 'including, making the event exist for the maximum number of interested parties, whether they are competent or incompetent' (Stengers, 2000: 103). So, to return to the case of Hirst, the links that attest to a world we live in as media are indeed 'accessible, in several alternative sequences, at the flick of a switch' (Williams, 1974: 94). Flicking through the pages of the Sunday colour supplement we can find out that David Beckham bought one of Hirst's hearts with dead butterflies for Posh for her birthday. Or, we can make a note to ourselves that 'for those with smaller budgets, signed photographic prints of a spot painting entitled Valium are being sold for $2,500 in an edition of 500 from Eyestorm.com' (*The Economist*).

But, on the other hand, what makes links in Hirst's trade mark style *is* what stands in the way, for the associations produced by the links marked by a brand name are subject to judgements of (intellectual property) law.[18] And here the terms of trade mark law become relevant. In very general terms, there has been an expansion in trade mark law in the past 20 or so years that involves a movement away from a 'confusion' definition of infringement (as to the origin of the product) towards a broader 'dilution' definition. The dilution definition of infringement precludes all unauthorized uses that would lessen (or take advantage of) a mark's distinctiveness.

Thus, it used to be the case that trade mark infringement would only be found where the use of a protected mark by someone (X) other than its owner (Y) was likely to cause consumers to be *confused* as to the origin of the product to which the mark was attached. The issue was whether consumers would think that X's product actually came from Y. Now it is increasingly being suggested – with varying degrees of success – that if X's use of Y's signs on its product causes consumers to be reminded of Y on seeing X's product, even while knowing that X and Y are distinct traders, infringement has occurred. In other words, *creating associations between products is becoming established as the exclusive prerogative of the trade mark owner*; associations created by other producers can be legally prevented if they *dilute* the first mark.[19] In other words, the recognition of the claim to be a brand name – the acceptance of a trade mark style – is a matter of the open, inclusive indeterminacy of switches, *and* the closed, exclusionary judgements of associations. In short, trade mark style is an open and shut box.

Coda

Let me offer a final, pharmaceutical example of the emergence of novelty in mixed media, that of InnoCentive.inc (www.innocentive.com).[20] This is an e-business company, founded in 2001 by Eli Lilly, a 'leading innovation-driven pharmaceutical company'. As described by Alph Bingham, who is said to be one of is founders, InnoCentive.inc is a 'talk-radio approach' to scientific R&D:

> InnoCentive is the first on-line scientific network created specifically for the global research and development (R&D) community ... With InnoCentive, global corporations have the opportunity to find solutions to tough R&D challenges by tapping into a worldwide network of uniquely prepared scientific minds. Leading corporations such as BASF, Dow Chemical, Eli Lilly, Procter & Gamble, Syngenta and others are posting their problems on the Inno-Centive site to take advantage of this worldwide network. Already, more than 25,000 leading scientists and researchers in 125 countries have registered as Solver scientists and are able to submit solutions to Challenges for potential financial reward. (Hussein, www.innocentive.com)

An InnoCentive Challenge is a scientific problem that is posted by a Seeker company; if a Solution is selected as 'best' by the Seeker Company, the Solver ('usually a scientist') receives a financial reward, which varies with the Challenge. InnoCentive.inc posts the Challenges along with their deadline or cut-off date, provides 'a secure space online' called a 'Project Room' for Solvers to view the details and requirements of a challenge and submit solution proposals, and requires both parties to sign an intellectual property agreement (in which the Solver 'typically relinquishes all rights'). Its own profits are derived from a percentage of the sum paid by the Seeker Company for the solution it deems 'best'. Challenges may include the identification of methods ('a quantitative method for analysis of a polymer', InnoCentive 1594674), and the provision of molecule libraries ('quantities of small peptide molecules that contain from 2 to 10 amino acids', InnoCentive 716103; see Barry, this issue) as well as the creation of new entities. Among the latter at the time of writing is a challenge to identify a 'novel additive with "greener properties"' (InnoCentive 1186758). The innovation incentive in this case is US$50,000, deadline 30 April 2004.

According to Hirst:

> I just don't understand it. People tell you what things are worth, but it's not what they're worth. I mean, you've got someone here who makes a product and then you've got a consumer. I just keep focused on that: you've got someone there and someone there. And as long as you're in one of those positions you're going to be alright. (In Burn and Hirst, 2001: 154)

Whether or not things will work out 'alright', whether or not the Inno-Centive.inc solution (to the problem of invention) is the 'best', the argument made here is that the conditions under which creativity, invention and discovery are recognized are being altered. The bringing of the world into existence as mixed media is being marked by brand names; and this transformation of the author function provides the conditions, as Stengers (2000) would put it, for a novel distribution of proof.

Acknowledgements

I would like to thank Mariam Fraser for her very helpful comments on an earlier version of this article. I would also like to thank Andrew Barry for information about the pharmaceutical industry.

Notes

1. According to Simon Ford (1996), the earliest usage of 'young British artists' was for the British Pavilion of the 1996 Venice Biennale. He also records other names for the grouping including the 'neo-conceptual bratpack', by Sarah Greenberg, *Artnews* September 1995, 'The Brit Pack', by Patricia Bickers, *The Brit Pack: Contemporary British Art, The View from Abroad*, 1995, and the 'Britpop' artists, by Waldemar Januszczak (*Sunday Times*, 3 December 1995).

2. Stengers does not use the term author function itself, but describes the emergence of 'a new use of reason centered on the question "Who is the author?"' (1997: 155).

3. Lev Manovich proposes that the 1960s saw the arrival of a new stage in the history of media, which he calls meta-media society. He writes:

> The new avant-garde is no longer concerned with seeing or representing the world in new ways but rather with accessing and using in new ways previously accumulated media. In this respect new media is post-media or meta-media, as it uses old media as its primary material. (2004)

4. Martin Maloney describes the spot paintings as 'luxury designer goods' (1997: 33).

5. This account of Williams' use of the notion of flow recognizes that Williams does not make the distinction that Deleuze draws between the rational and irrational

interval (1986, 1989; see also Rodowick, 1997). This is a distinction between the interval that produces association and the interval as interstice or cut. Deleuze draws this distinction most clearly in a discussion of a Jean-Luc Godard film, *Je t'aime je t'aime*. He writes,

> The so-called classical cinema works above all through linkage of images, and subordinates cuts to this linkage. On the mathematical analogy, the cuts which divide up two series of images are rational, in the sense that they constitute either the final image of the first series, or the first image of the second. This is the case of the 'dissolve' in its various forms . . . Now, modern cinema can communicate with the old, and the distinction between the two can be very relative. However, it will be defined ideally by a reversal where the image is unlinked and the cut begins to have an importance in itself. The cut, or interstice, between two series of images no longer forms part of either of the two series: it is the equivalent of an irrational cut, which determines the non-commensurable relations between images . . . each shot is deframed in relation to the framing of the following shot. (1989: 213–14)

The usefulness of this distinction in relation to cinema is clear; however, as Deleuze notes it is a relative distinction. The suggestion here is that the brand may be differentiated by both rational and irrational intervals.

6. This is the name of an exhibition held in Paris in 1996, curated by Hans Ulrich Oberist, featuring work from many of the artists who comprise yBas, including Hirst.

7. Many commentators suggest that Hirst is primarily concerned with death, but the argument outlined here is that, as he says, 'I think I've got an obsession with death, but I think it's like a celebration of life rather than something morbid' (in Burn and Hirst, 2001: 21).

8. The colours of the spot paintings are similarly said by Hirst to be inspired by the colours used in product catalogues for commercial drug firms; and many of the titles of spot paintings are the names of drugs.

9. As Hirst says:

> I want the world to be solid. But it's so fluid and there's absolutely nothing you can do about it. I mean I'm absolutely not interested in tying things down. Or having control, to any major extent. It's like: this is the way I make it because this is the way things are working now, but I don't expect them to be working like that tomorrow. Suddenly, like, Silk Cut can go off the market. Mercury gets into Silk Cut cigarettes and a thousand people die and there's nothing you can do about it. Boom! Gone. A worldwide brand. But I enjoy that. I don't want to tie the world down. (In Burn and Hirst, 2001: 37)

10. In an interview, Hirst says: 'In physics, you know, if they can't find the answer they want, they change the question. As long as you're prepared to do that, or change your way of looking at the question, there's nothing you can't do' (in Burn and Hirst, 2001: 27).

11. In an interview, Hirst remarks:

> I once shared a studio with Angus (Fairhurst), and I was doing a painting in the corner one morning and he suddenly goes, *'Don't!'* That's why you don't

share studios with people. '*Don't! You'll ruin it!*' And then going, 'Nothing', afterwards. 'Sorry, it was nothing.' Trying to give you your space back. (Burn, 2000: 205)

12. In what I think is only apparent contradistinction, Stengers writes:

> A scientist is never alone in his laboratory, like an isolatable subject. His laboratory, like his texts, like his representations, is populated with references not only to those who could put them in question, but also to all those for whom they could make a difference. How did Pasteur represent a microbe? As Bruno Latour writes, 'this new microscopic being is both anti-Liebig (the ferments are alive) and anti-Pouchet (they are not born spontaneously).' But Pasteur had already envisioned other possible significations, and numerous other practices where his microbes could make a difference. (2000: 95)

13. Caroline Jones explains Warhol's factory as an alternative to 'The dominant topos of the American artist . . . a solitary (male) genius, alone in his studio, sole witness to the miraculous creation of art' (quoted in Alpers, 1998: 403).

14. This transformation in the space of the artist's workplace is in part a consequence of the proliferation of museums, galleries, and art publications in the latter part of the 20th century (Schubert, 2000) but it is also a result of the speed and intensity of the flows of mediation – and immediacy – that characterize contemporary culture. Buskirk's discussion of the conditions contributing to the rise of trade mark styles includes the rise of the celebrity artist within the art world and the routine appropriation of mass-produced cultural products by artists. But it does not include a serious consideration of the significance of the reverse movement – that is, the appropriation of art work in popular culture, other than to suggest that such movements must involve a degree of naivete on the part of advertisers and such like. Thus she says, 'But subtle or implied criticism is something that advertisers and others in the mass media are capable of overlooking. This tendency is demonstrated by the number of advertising endorsements that Warhol was asked to do' (1992: 108). But it is increasingly clear that such reverse movements – that is, the use, naive or otherwise, of artistic works in mass cultural forms – have to be acknowledged. Such two-way movements are not simply incidental but actually constitutive of the art world. This is especially clear in relation to the circulation and valuation of Hirst's work because of the patronage of Charles Saatchi, whose reputation in the advertising world preceded his entry into the art world as a collector. While Saatchi's motives in the purchase and display of the work of Hirst and other yBas are the focus of much discussion in the art world (Hirst describes himself as 'a conduit from art to money' [in Burn and Hirst, 2001: 192]), it is clear that its boundaries are not simply leaky but functionally permeable.

15. Lev Manovich also draws a parallel between white cubes and black boxes, but in his case the black box is the cinema.

16. Stengers argues that computer simulations are transforming the status of what science calls a model, complicating the links between science and fiction. She writes:

> For a long time the 'model' has signified, in scientific practices, a 'poetic' activity that is incapable of making its author a judge. The model has an author who knows he cannot claim to be forgotten. But the model simulated

on a computer introduces such a distance between the author's hypotheses and the engendered behaviour that the author speaks of the computer in the same way as the experimenter speaks of the phenomenon: as if, adequately interrogated, it can act as the authority. 'The simulation shows that . . .' is a statement that, henceforth, sometimes takes the role of 'the experiment shows that . . .' Here there are new histories beginning, with new types of authors, new stakes, and new controversies. (1997: 168)

17. Stengers also defines *inter-esse* as 'the sensibility to a possible becoming' (2000: 92).

18. Hirst has not in fact registered his name as a trade mark; that is, he has not sought protection in law for the exclusive use of the name Hirst. At present, his name does not actually identify a trade mark style in law.

19. And here the very use of the term 'dilution' seems to indicate a tacit legal acknowledgement that what is at issue here is the organization of the logics of flow.

20. I would like to thank Mariam Fraser for pointing out this venture to me.

References

Alpers, S. (1998) 'The Studio, the Laboratory, and the Vexations of Art', pp. 401–17 in C.A. Jones and P. Galison (eds) *Picturing Science, Producing Art*. London: Routledge.

Arlidge. J. (2000) 'Britannia's Brand New Start', *Observer*, Sunday, 14 May.

Barry, A., 'Pharmaceutical Matters: The Invention of Informed Materials', *Theory, Culture & Society* 22(1): 51–69.

Burn, G. (2000) 'The Knives Are Out', *The Guardian* 10 April.

Burn, G. and D. Hirst (2001) *On the Way to Work*. London: Faber and Faber.

Buskirk, M. (1992) 'Commodification as Censor: Copyright and Fair Use', *October* 60: 82–109.

Corris, M. (1992) 'Damien Hirst', *Artforum* January: 96.

Deleuze, G. (1986) *Cinema 1: The Movement Image*. London: Athlone Press.

Deleuze, G. (1989) *Cinema 2: The Time Image*. London: Athlone Press.

Dyer, C. (2000) 'Hirst Pays Up for Hymn that Wasn't His', *The Guardian* 19 May.

The Economist (2001) 'Portrait of the Artist as a Brand', 10 February.

Ferrier, H. (2001) 'Successful Switch Strategies', pp. 60–84 in T. Blackett and R. Robins (eds) *Brand Medicine: The Role of Branding in the Pharmaceutical Industry*. Basingstoke: Palgrave.

Ford, S. (1996) 'Myth Making', *Art Monthly* 3: 3–9.

Foucault, M. ([1969] 1998) 'What is an Author?', pp. 187–204 in M. Foucault *Aesthetics, Method and Epistemology*, ed. J. Faubion. London: Penguin.

Frow, J. (2002) 'Signature and Brand', pp. 56–74 in J. Collins (ed.) *High-Pop: Making Culture into Popular Entertainment*. Oxford: Blackwell.

Gandelman, C. (1985) 'The Semiotics of Signatures in Paintings: A Peircian Analysis', *American Journal of Semiotics* 3: 73–108.

Hirst, D. (1995) ' "Brilliant!" New Art from London', interview by Marcelo Spinelli.

Hirst, D. (1996) 'Damien Hirst: No Sense of Absolute Corruption', interview by Stuart Morgan, New York: Gagosian Gallery.

Jones, J. (2000) 'There Goes Art's Last Hope', *The Guardian* 30 November.

Krauss, R. (1999) *'A Voyage on the North Sea': Art in the Age of the Post-Medium Condition*. London: Thames and Hudson.

Latour, B. (1987) *Science in Action: How to Follow Scientists and Engineers through Society*. Cambridge, MA: Harvard University Press.

Latour, B. (1996) 'On Interobjectivity', *Mind, Culture and Activity* 3(4): 228–45.

Leith, W. (1999) 'Avoiding the Sharks', *The Observer* 14 February.

Lury, C. (2004) *Brands: The Logos of the Global Economy*. London: Routledge.

Maloney, M. (1997) 'Everyone a Winner! Selected British Art from the Saatchi Collection', in N. Rosenthal and S. Fraquelli (eds) *Sensation: Young British Artists from the Saatchi Collection*. London: Thames and Hudson.

Manovich, L. (2003) 'The Poetics of Augmented Space', l.manovich.net, (12 March).

Manovich, L. (2004) 'Avant-Garde as Software', l.manovich.net (30 April).

Massumi, B. (2000) 'Too-Blue: Colour Patch for an Expanded Empiricism', *Cultural Studies* 14(2): 177–226.

McLuhan, M. (1996) *Understanding Media: The Extensions of Man*. London: Routledge.

Moor, L. (2003) 'Branded Spaces: The Mediation of Commercial Forms', unpublished PhD thesis, University of London.

O'Doherty, B. (2000) *Inside the White Cube: The Ideology of the Gallery Space*. Berkeley: University of California Press.

Rabinow, P. (1998) 'Secede and Assemble: Ready-Made Events in Molecular Biology', pp: 151–79 in T. Yamomoto (ed.) *Philosophical Designs for a Socio-Cultural Transformation Beyond Violence and the Modern Era*. Boulder, CO: Rowman and Littlefield.

Rodowick, D.N. (1997) *Gilles Deleuze's Time Machine*. Durham, NC: Duke University Press.

Schubert, K. (2000) *The Curator's Egg: The Evolution of the Museum Concept from the French Revolution to the Present Day*. London: One-Off Press.

Simon, H.A. ([1969] 1981) *The Sciences of the Artificial*. Cambridge, MA: MIT Press.

Stengers, I. (1997) 'Who is an Author?', pp. 153–76 in *Power and Invention: Situating Science*. Minneapolis: University of Minnesota Press.

Stengers, I. (2000) *The Invention of Modern Science*. Minneapolis: University of Minnesota Press.

Williams, R. (1974) *Television: Technology and Cultural Form*. London: Collins.

Wilson, A. (1994) 'Out of Control', *Art Monthly* 177: 3–9.

Celia Lury is Professor of Sociology at Goldsmiths College. She has recently completed a book on *Brands: The Logos of the Global Economy* (Routledge, 2004), and is working on a book, *The Global Culture Industry*, with Scott Lash. She is a member of the Centre for the Study of Invention and Social Process at Goldsmiths.

The New Economy, Property and Personhood

Lisa Adkins

Introduction

THE FOCUS of this article is the new economy, or what some have termed a virtual, reflexive or network economy. While the concept of the new economy is often critiqued for its vagueness – an opaqueness which diverse conceptual and more empirically based definitions have contributed towards – this article aims to identify how in economic arrangements characterized by knowledge and service intensity, relations of property, and especially the relations between people and their labour, are being reworked. In so doing, this article therefore aims to contribute to the project of specifying the dynamics of the new economy (Callon et al., 2002; Castells, 1996; Lury, 2003; Thrift, 1998). Its point of departure is the apparent contradiction, often made visible in studies and analyses of the new economy, that as the economy becomes more and more virtual (Carrier and Miller, 1998), reflexive (Thrift, 1998), networked (Castells, 1996, 2000) or, as it is sometimes put, immaterial (Hardt and Negri, 2000), a greater emphasis is being placed on issues of embodied performance and the significance of human or physical capital appears to be intensifying. In pointing to this contradiction I am certainly not meaning to suggest that the new economy involves a kind of retreat from the material or the physical. Studies of the new economy have done enough to dispel this myth and indeed have shown how material relations are being reworked in this context, for instance, how the replacement of the accumulation capital with the accumulation of information reworks power and property as largely informational (Lash, 2002; Rodowick, 2001). However, while this is so, what studies of the new economy have tended to leave untouched is a consideration of how material relations may be reconstituting vis-à-vis the people who may be working in the new economy. Therefore, while such analyses may show how new technologies of communication, or the shifting characteristics of goods and products reconstitute material relations, for example, how various forms of data copyrighting are reworking the material processes that constitute public and private life (Haraway, 1997), most analyses of the new economy tend to stop short of considering how this reworking of materiality works out in regard to people. This is not to say that analysts of the new economy do not discuss people – for they do. But when they do so, a version of personhood tends to be invoked which side-steps a consideration of how personhood itself may be materially reconstituting in the new economy. Specifically, when people are discussed, they are assumed to be largely in control of and indeed to own their own identities and bodies – a version of personhood which I shall term in this article, following the work of Carol Pateman (1988), a social contract model. One expression of this view of personhood in recent literature on the new economy is found in the emphasis on the aforementioned notions of human, physical or embodied capital. Deriving very loosely from the work of Pierre Bourdieu, such conceptions are widely invoked to get at the ways in which the significance of human capital appears to be intensifying in the new economy. What the concept of human or embodied capital assumes, however, is that people can own or at the very least accumulate forms of capital: that various forms of capital stick to the human subject, a version of personhood which assumes that subjects may own property in the person and may abstract or disentangle that property and trade it as a resource for exchange. What I shall suggest in this article, however, is that in the new economy people cannot unproblematically claim to own and straightforwardly accumulate property in the person, since the relations between property and people are being restructured. More particularly, I will attempt to illuminate how in the new economy qualities previously associated with people are being disentangled, are the object of processes of qualification and re-qualification, and moreover how claims to these qualities are made not through claims towards ownership of these qualities as forms of property in the person (as labour power), but rather through claims which operate external to the domain of personhood.

The Framing of the New Economy

While writings on the new economy are legion, nonetheless there tends to be an explicit and/or implicit agreement within them that economic action is in the process of fundamental revision. What characterizes economic action is no longer its sociological features – for instance, innovation around systems of generative rule-resource sets, involving a continuous production and reproduction of society or social systems, including the production and reproduction of class and gender and other classic social formations of modernity – but rather its relative openness and flexibility, an openness which may be understood to be linked to a retrocession of social structures in contemporary societies (Beck et al., 1994), the emergence of an information or network society (Castells, 1996, 2000) and post-social modes of sociality (Knorr-Cetina and Bruegger, 2000, 2002). Substantive studies of the new economy attest to this more open character of action. Such accounts stress the retreat of socio-structural formations which previously framed possibilities for action (for instance, those of occupation and organization); an intensification of self-regulation and self-reflexivity; how working practices are increasingly networking practices; a proliferating significance of intermediaries; and how workers tend not to share narratives or life histories and how continuous effort and emphasis are placed on working at making and re-making networks and relationships – people do not 'have' relationships, instead there is a focus on doing relationships and on relationship management (Adkins, 2002; Casey, 1995; Castells, 1996, 2000; Knorr-Cetina, 2004; Lash, 1994; Lury, 2003; Martin, 1994, 2000; Sennett, 1998; Wittel, 2001). Indeed, in such accounts of the new economy it has been stressed that the significance of factors which in modernity were paradigmatic of the framing of economic action and (especially) economic exchange recedes. The significance of contracts, for example (as well as products), retreats as emphasis is placed on doing, making and remaking relationships, a mode of action which has been termed a network sociality (Wittel, 2001).

While not explicitly concerned with the new economy, nonetheless the theoretical resources offered by Callon (1998a, 1998b; Callon et al., 2002) in his recent writings on economic activity appear to have particular purchase for understanding this more open, flexible character of economic action characteristic of the new economy.[1] Very briefly put, Callon's argument is that Goffman's notion of framing may be usefully applied to economic activity. Framing concerns the idea that interaction takes place – or perhaps more correctly said, can only take place – within framed settings, that is, where a boundary is established (but by no means abolished) between an event and the outside world. A theatre performance is, for example, only made possible via the fact that everyone involved (from actors through usherettes to audience members) agrees on a particular framing of the event. Indeed, such a performance may only take place via a tacit agreement between all parties. The audience knows, for instance, what rules to obey, that they should fall silent when the curtain rises and so on. Framing thus both enables and stabilizes action (at least temporarily) since mechanisms of framing produce an implicit agreement on the character and rules pertaining to action. Framing is therefore akin to Bourdieu's notion of habitus – it both generates and shapes action – and it is this notion of framing which Callon applies to economic activity. The negotiation of a contract or the performance (or what at times Callon refers to as per-formatting) of a commercial transaction, for example, presupposes:

> . . . a framing of the action without which it would be impossible to reach an agreement, in the same way that in order to play a game of chess, two players must agree to submit to the rules and sit down at a chessboard which physically circumscribes the world within which the action will take place. (Callon, 1998b: 250)

In other words, economic calculations can only be performed and completed if the agents and goods involved in these calculations are, to use Callon's terms, disentangled and framed (Callon, 1998a: 16). However, Callon by no means reduces framing to contracts. Indeed, he argues that the framing of the economic activity cannot be achieved by contract alone. Callon discusses this in terms of a paradox: 'a totally successful frame would condemn the contract to the sterile reiteration of existing knowledge' (1998b: 255). This paradox arises because while a contract cannot be framed, stabilized and fulfilled without the participation of a range of heterogeneous elements – equipment, objects and specialists – at the same time, this very participation leads to externalities or what Callon calls overflows. That is, while such linked elements frame the contract and the performance, they take part in its overflowing, and it is precisely this overflowing which makes the *contract productive*. Researchers performing a contract, for example, interact with colleagues at conferences, may move temporarily to different research laboratories and may publish research findings which provoke debate. In short, the very elements that stabilize a contract are sources of overflow (since frames are never entirely closed off or sealed), and it is this overflowing that produces *productivity*. Overflowing may thus be thought of as being part of the production of production (Adkins and Lury, 1999). The different elements that frame a contract should therefore not simply be regarded as resources but also as *intermediaries* since they both frame interactions and present openings into wider networks to which they give access. Callon's paradox regarding contract may then be resolved by rethinking action. Action does not have a source, for example, the contract, but rather is always mediated, to use Callon's words 'what counts are the mediations and not the contracts' (1998b: 267).

What is significant about this analysis of economic activity is that it captures the relative openness of economic action, that is, the very openness and flexibility which have been attributed to action in the new economy. Indeed, Callon's economic activity is constituted in a (socio-technical) 'network sociality'. Callon's account therefore provides resources to rethink economic action outside of traditional registers, including those which understand economic activity as driven by the logic of structure and agency and rule resource sets. Moreover, Callon's account explicitly underscores how and why economic action requires such rethinking – economic action is not framed or constituted by (relatively) closed or fixed sources but is found in open, moving, heterogeneous networks.

Yet while Callon may provide resources for the theorization of action in the new economy, nonetheless accounts of the new economy are certainly not without their critics. One of the most widely rehearsed criticisms of this body of work is that analyses of the new economy risk dissolving economic processes into culture, communications and information and in so doing overlook what distinguishes the economy from other fields. Thus, and in a critique of purely culturalist explanations of economic activity, Slater (2002a) has suggested that such analyses have a tendency to play down the fact that exchange in the economic field concerns the alienation of goods in the form of property. Indeed, Slater claims that such analyses may overlook the most fundamental framing of economy, namely specific forms of owner-ship and property right presumed in commodity exchange. He writes:

> What distinguishes market transactions from non-market transactions is *alienation* . . . Market exchange is commodity exchange. It presumes a form of property right in which a transfer of ownership ends all claims of the previous owner: the object is thoroughly alienated. When we finish the trans-action, we are quits. (Slater, 2002a: 237, emphasis in original)

And he goes on:

> There really is such a thing called the market because there is such a thing as commodity exchange which is characterized by specific forms of property right. This form of contract means that exchange is disembedded in seriously consequential ways. (Slater, 2002a: 240)

In this view, it is property rights which lie behind disentanglement – the key process which stabilizes products, detaches them from the networks from which they originate and allows for exchange – and which should be the proper focus of studies of the contemporary economy. Thus, while studies of the new economy tell us that issues of contract are increasingly less significant, Slater would have us see the classic market exchange involving specific forms of property rights and ownership as the fundamental framing of the economy.

But is this so? I want to suggest that, in the new economy, property right and ownership are being restructured in some significant ways. In particular, it is my contention that property and ownership issues in the new economy are not simply tied into the classic market exchange as outlined by Slater. This is so, I will suggest, not only because the ownership and property rights at issue in the new economy are those relating to contem-porary forms of *authorship* – a form of property relation where it is never entirely clear, to paraphrase Slater, exactly when we are quits – but also because this form of property relation involves a fundamental reworking of the relationship between property and the person. This latter, I will argue further, involves what some prefer to understand, I think rather mislead-ingly, as the commodification or commercialization of social relations, but what I will suggest is better described as part of the wide-ranging process of the patenting of matter previously coded as natural and/or social – a process which has been described as type or kind becoming brand (Haraway, 1997; Lury, 2000). This process displaces relations of ownership relating to property and commodity exchange, that is, the thoroughgoing alienation or disentanglement of objects. To illustrate this argument I will discuss labour, a difficult subject to think about, in terms of the debates I am engaging with in this article, since of course it has never been entirely clear that labour can be fully alienated or disentangled at all. Yet labour is perhaps one of the most interesting aspects of the new economy, since it is here we witness paradigmatically the restructuring of the relationship between property and people that this article seeks to underscore.

The Social Contract and Property in the Person

To begin my argument it is important to establish the characteristics of the organization of the relations between property, ownership and people in industrial capitalism, characteristics which are illustrated well by Carol Pateman (1988) in her work on the social contract. As is well known, Pateman's analysis was a critique of the social contract, that is, the emergence of the social. She argued that the liberal claim that political right is created through a social contract – that free social relations take a contractual form – obscures the ways in which liberal contracts involve a double movement of both freedom and subjection.[2] What is important for the purposes of my argument here is Pateman's focus on property. As she points out, for the most part, discussions of property in regard to contract focus on property as either material goods, land and capital, or as the interest that individuals can be said to have in civil freedom (Pateman, 1988: 5). However, this focus tends to overlook a form of property that is the subject of contract: the property that individuals are held to own in their persons, the very form of property which the social contract declares constitutes people as free individuals. The employment contract exemplifies this aspect of contract well, as it assumes that employment involves the contracting out of property in the person, that is, skills, attributes and capacities (in short, labour power) in exchange for a wage. Pateman argues, however, that the claim that labour power is contracted out, not labour, bodies or persons, enables proponents of contract to claim that the employment contract constitutes a free relation. This claim, Pateman goes on, is a political fiction, since it negates the way in which the employment contract is not a free exchange involving the right to alienate property in the person. This fiction can only be accepted if it is assumed that abilities can acquire an external relation to an individual, and can be treated as if they are property, that is, that abilities, capacities and skills can be alienated or disentangled, and that it is this process of disentanglement, together with the workers' ownership of these skills and abilities, which defines a worker as a free individual. Thus, this fiction can only be accepted if it is taken for granted that the individual owns his or her labour power (body and capacities) as a commodity, and that this relation to labour power is the same relation enacted in regard to any other form of material property.

The difficulty with this assumption, however, and especially the assumption that individuals own property in their persons, is, as Pateman argues, that labour is not the same as other forms of material property. Unlike other commodities, and again to think with the language of Slater, in the case of labour power, being quits is particularly difficult, since it cannot be straightforwardly disentangled from the person. As Pateman puts it: 'The worker's capacities are developed over time and they form an integral part of his self-esteem and self-identity; capacities are *internally* not externally related to the person' (1988: 150, emphasis added). Labour power therefore always requires the presence of its owner and hence cannot be disentangled like other forms of property. For Pateman, therefore, the subject of the employment contract is not labour power, but the worker and his/her labour. That is, because a worker cannot be separated from his/her capacities – which in Pateman's vision are accumulated over time and become, as Bourdieu might have it, part of the habitus of the worker – what is sold is command over the use of his/her body and him/herself. As a consequence, only some workers can achieve the ideal of ownership of property in the person as aspired towards in the employment contract. Indeed, Pateman argued that this ideal is only ever reached by a few men since an institutionalized system of socio-legal rights afforded only certain men such jurisdiction over property in the person. We may think of the artisan as the optimal figure who embodied such rights, for whom at a certain historical moment employment associations and state legislation (from employment through housing to family policy) ensured that he could claim to own and contract out the skills he accumulated over his years of apprenticeship and time spent on the job.

The exclusions from this are obvious. The figure of the artisan always, for example, assumes its other: the figure of the housewife. Pateman argued that the employment contract presupposed the marriage contract, and in particular a housewife who took care of the worker's daily needs. In the social contract, a housewife, Pateman argued, is a sexual subject who lacks jurisdiction over the property in her person, which included labour power. Thus, in conditions of the social contract, 'women' could not be workers in the same sense as men as they lacked the right to alienate property in the person. Or perhaps more precisely, in the social contract the socio-technical devices which made such forms of alienation possible were not fully operative for women.

Authorship: From Property in the Person to Audience Effect

Pateman has recently argued (2002) that these ideas regarding labour, property and persons are applicable to the new economy and in particular to globalization, citizenship and employment. However, I am not so sure that employment in the new economy can be characterized as involving struggles over rights to property in the person. This, I think, is evidenced in the ways in which what Pateman terms capacities and abilities cannot be unproblematically accumulated by workers in the new economy, since such qualities are no longer figured in terms of an internal relation to the person. Indeed, in the new economy what is at issue in terms of workers is not ownership of property in the person, since as I alluded to earlier, in the new economy property rights regarding labour are more akin to those which characterize contemporary forms of authorship. Now authorship has a long history of debate particularly when it comes to cultural production and reproduction, that is to creative labour, but nonetheless what is important regarding these debates in terms of the concerns of this article are recent accounts of shifts in regimes of cultural production (that is, the production of culture for economic exchange), which describe a move from a regime of cultural production ordered by authorship, originality and signature to one ordered by the brand, branding and simulation (Frow, 2002; Lury, 1993). While a complex set of changes in aesthetic and institutional arrangements accompany this shift, my interest here is with how such accounts suggest that the conditions of authorship, and especially the relations between the creative labour process and ownership of that labour may be changing (albeit unevenly).

As a range of accounts of authorship have made clear, the historical emergence of the notion of the author was bound up with the conceptions of the individual, self-possession and subjectivity. Thus the author as historical figure was defined as a creative, autonomous individual. But authorship defined a lot more than a specific kind of (and special) individual. It also established cultural works as a form of property since the notion of authorship defined the distinctiveness and value of a cultural work in terms of what the creator of cultural works already owned – his creative will or personality. Therefore, under conditions of cultural production ordered by the notion of authorship, the creative will or personality of the cultural producer was figured as embodied in cultural works. In short, and as Lury has put it 'it was through the author-function that cultural value became a thing, a product and a possession caught in a circuit of property values' (1993: 23). As an institutionally established set of links between the creative labour process and creative works, authorship thus not only distinguished creative labour from other forms of labour but moreover defined creative labour vis-à-vis creative works as (and to use Pateman's terms) a particular kind of property in the person.

While it is widely accepted that the cultural and commercial authority of the idea of the author as originator in cultural production and reproduction has declined, nonetheless Lury (1993) has suggested that this should not necessarily be read to mean that the status and conditions of creative labour have been entirely flattened out, that is, that they are the same as those for other forms of labour. Lury argues instead that these differences may in fact be maintained by less direct means. Specifically, creative labour may be able to retain privileged conditions via claims regarding the commercial effects of a cultural good. Thus, Lury points to a number of mechanisms operative in the contemporary culture industries through which creative labour may win recognition for such effects, which crucially include the audience's ability to decipher the signs of creative labour in the cultural product. Lury notes how a number of techniques go towards making such effects visible, which most notably include the techniques and procedures associated with audience ratings. Despite the demise of the author-function, creative labour may therefore be able to win recognition (and hence claims to ownership of that labour) via greater attention to the activities of the audience. In short, what Lury's account underscores is a shift away from definitions of cultural value and distinctiveness based on notions of the embodiment of the creative will or expression of the cultural producer in cultural goods towards definitions of value based on processes of reception or what she terms (following Benjamin, 1970) exhibition value. Indeed, more generally, Lury points out that in the contemporary culture industries one of the surest guarantees of success is the strategic linking of the signs of creative labour with the exhibition value of works by stars or personalities via the self-conscious or reflexive management of appearances, publicity and promotion.

What is important regarding such accounts of changes to regimes of cultural production and reproduction in terms of my concerns here (although not necessarily made explicit) is that they highlight for the case of creative labour a reworking of the relations between persons and the ownership of labour.[3] Specifically, in regimes of cultural production organized by the principle of the author-function, creative labour is figured (via a range of socio-technical devices) as a property of the person, while in regimes of cultural production organized via branding and exhibition value, creative labour is defined (again via a variety of socio-technical devices), not in terms of the relations of personhood but in relations external to the person. Thus ownership of that labour is made less in terms of claims concerning the creativity and uniqueness (that is the capacities and abilities) of the cultural producer (the individual) and more in terms of process of reception, that is, in terms of the commercial effects of cultural goods vis-à-vis the intended audience.

The New Economy, Authorship and the Audience

While the accounts of shifts in the relations of property (rights to the ownership of labour) and authorship I highlighted earlier relate specifically to creative labour, it is my suggestion that the shift in the relationship between property and the person vis-à-vis authorship is also evident in regard to labour in the new economy. Put differently, labour in the new economy is taking on the characteristics associated with contemporary creative labour.[4] This point is particularly well illustrated (although perhaps rather counter intuitively) in relation to gender. In the social contract a range of socio-technical devices – or what may be termed technologies of inscription – defined gender not just simply as an embodied quality of the person, but as a taxonomic type coded as natural and/or social. Thus, in the social contract, gender was not alienable as a form of property. However, in the new economy characterized by data, knowledge and service intensity, gender is detachable from the person and indeed may be alienable as a form of (cultural) property. This is indicated perhaps most clearly in studies tracking the emergence of the ideals of flexibility and adaptability in corporate cultures, and especially those which have underscored how flexibility has emerged as a new ideal for employees (Brown, 2003; Martin, 1994, 2000). Here it has been emphasized how a range of techniques, particularly those mobilized in ceaseless training and assessment programmes, performance based pay systems, and continuing educational programmes figure mobility, flexibility and adaptability as desirable characteristics of workers (Brown, 2003; Hinchliffe, 2000).[5] Such techniques displace and replace those socio-technical mechanisms which worked to constitute traditional workplace hierarchies, including those running along (and inscribing) axes of gender. Indeed, such techniques may be understood to destabilize the social contract model of gender as they work to undo the internal link between gender and persons. While not explicating such mechanisms in these terms, nonetheless Martin (1994) has described a range of techniques which may be understood to be performing such a destabilization. Specifically, she describes training exercises:

> . . . in which teams of men and women workers and managers of all ages and physiques . . . climbed forty-foot towers and leaped off into space on a zip line, climbed forty-foot high walls and rappelled down again, climbed a twenty-five-foot high telephone pole, which wobbled, stood up on a twelve-inch platform on the top, which swiveled, turned around 180 degrees, and again leaped off into space. (Martin, 1994: 212)

Such workplace exercises are designed to break down traditional workplace hierarchies such as those between management and labour, men and women. Indeed, such exercises are intended

> to *scramble* the characteristics usually associated with males and females. Men can feel fear on the high ropes course and thereby learn to express their vulnerability; women can feel brave and thereby learn to see their ability to lead. Men and women can also learn to appreciate these unaccustomed capacities in each other. (Martin, 1994: 213, emphasis added)

Such scrambling exercises suggest that in contrast to the social contract where gender was organized as an embodied quality of persons figured in terms of relations of internality, in the new economy, gender has become detached from the person and is a mobile object (Adkins, 2001, 2002). Thus such training practices do not assume that gender is a signifier of essential sex nor a characteristic that workers simply possess by virtue of 'natural', 'social' or some other form of determination. Rather, gender is denaturalized and established as mobile, fluid, and indeterminate. Moreover, many management specialists Martin quotes refer to the necessity of 'scrambling' sexual difference for corporate success and growth. Thus one specialist recommends that corporate growth is dependent on '*sexual fusion and recombination*' involving a recognition and integration of 'those parts of ourselves – *male/female identities*, left/right brains, divergent/convergent thinking – that we have all neglected while striving for a strong sense of "self"' (Martin, 1994: 208, emphases added).

There is a range of implications of this disentanglement of gender from the person, but here I will concentrate on just two, which I see as being of particular significance. First, as implied by the quote from the management specialist established above, and in contrast to the sexual contract, in the new economy, gender becomes a form of property which can be exchanged, that is, gender is alienable from the person. This is illustrated well, and again to turn to recent studies of the new economy, not only in the ways that such studies underscore a reworking of gender from (relatively) fixed quality or stable type towards a mobile and flexible object, but also in the way they suggest gender in the new economy is increasingly figured as a workplace resource, that is, a form of capital which may be exchanged (Gray, 2003; Lovell, 2000), indeed, as an object of innovation (Adkins, 2000). Gray (2003), for example, has suggested that the figuring of gender as flexible and adaptable enables a deployment of gender as particular workplace skills and resources, especially aesthetic and stylistic skills. Put differently, the disentanglement of gender from people establishes gender as an alienable form of property with cultural value.

The reconfiguration of gender from type or kind to disentangled (and alienable) cultural object has been noted implicitly, or perhaps more correctly, sideways, in studies of the new economy.[6] This takes expression in claims that social relations in the new economy are increasingly commodified or commercialized; that a spatial and temporal co-presence of producers and consumers opens out previously hidden aspects of employee identity to economic transactions; that the boundaries between production and consumption are increasingly blurred; that action in the new economy (as already mentioned) no longer concerns rules and resources but is rather more open; that gender in the new economy is governed and is characterized by principles of performance, including the idea that such performances may secure workplace rewards (hence the claim that masculinity, femininity, and gender hybrids may be performed, mobilized, and contested by workers in a variety of ways in order to innovate and succeed in flexible corporations); and that gender (as well as other modern identities) in regard to the economy is increasingly subject to a process of aestheticization or stylization.

However, while a range of recent studies have variously noted these points, nonetheless they tend not to register their significance. For instance, they tend to argue in broad terms that these shifts are illustrative of the aestheticization of everyday life and/or a dedifferentiation of culture and economy and leave untouched how the shift from type or kind to cultural product involves a major reconfiguration of the relationship between property and the person in regard to gender. Indeed, such a consideration is rendered impossible since most studies of the new economy assume a social contract model of personhood where a person is assumed to own and to be in control of their identities and bodies. Thus, as Adkins and Lury (1999) have illustrated, studies of the new economy tend to simply assume that people working in the new economy own and/or are in control of performances of gender. Put differently, such studies adopt what might be thought of as an author-function approach to gender, whereby workers are viewed as the originators of gender performances (that is of cultural production) and moreover that such performances are a particular kind of property in the person.[7] Thus, there is a widespread assumption in studies of the new economy that gender is a cultural work which workers can unproblematically own, accumulate and exchange in return for workplace rewards.

Yet what this set of assumptions negates, and this is the second implication of the disentanglement of gender from the person which I wish to highlight in this article, is the way in which the shift from type or kind to cultural product reworks the relations between property and the person. Indeed, in the new economy, rather than by an author-function which locates gender as a cultural work as the property of the person, authorship vis-à-vis gender, like other forms of contemporary creative works, can only be claimed by less direct means. For as is the case for contemporary creative labour, authorship is only recognized vis-à-vis performances of gender via claims regarding the effects of the cultural product in regard to the intended audience.

This point is underscored in a range of studies of the new economy, from the earliest studies of service labour to the more recent studies of the network economy. Here it has been generally noted that employee effectivity is measured not, for example, in terms of units of production or quality of products, but in terms which relate to customers (see, e.g., Adkins, 1995; Beynon et al., 2002; Halford et al., 1997; Hochschild, 1983; Wittel, 2001).[8] Thus, in the new economy, customer satisfaction, appeased customers and customer loyalty are all key indicators of employee performance and a range of socio-technical devices such as the customer audit, customer benchmarking, customer surveys, customer focus groups and job descriptions and job training schemes which foreground customer care, comfort, pleasure and contentment are in place both to make visible and to measure such audience effects. Indeed, it has been noted that in the new economy worker subjectivity is significantly orientated towards the needs and desires of consumers (Du Gay, 1996), a point attested to by employee self-help manuals which advise employees of the significance of audience effect. In *The Brand You*, for example, Tom Peters (1999) makes this point explicit. Advising flexible employees to brand themselves to achieve the twin goals of differentiation and dramaticization, Peters notes that workers should always remember that what counts is *audience perception*. In short, in the new economy, the distance between production, labour and products and customers is increasing emptied out. This process may be understood as central to the reworking of economic action in the new economy. Specifically, the lack of distance between products and customers may be understood as contributing to the more open character of economic action, since rather than as spectators or simple consumers of cultural products, customers are now 'interactants' in cultural production.[9] Put differently, in the new economy the customer is a key intermediary in the framing of economic action, a point attested towards in the emergence of calculations of economic value (for example, by accountants and management consultants) which take place not with reference to physical or conventional capital, but to what are sometimes termed the 'intangibles' or intellectual capital which includes measures of customer loyalty and customer satisfaction, made visible by devices such as customer satisfaction indexes.

What is particularly interesting in terms of the argument I am putting forward here is that the significance of audience effect is particularly evident in relation to gender. For example, while in the new economy it is widely claimed that the mobilization of performances of femininity (for all workers) may be a workplace resource (see, e.g., Gray, 2003; Lovell, 2000), this is so only when such performances are recognized as having customer effects. Thus, emotionally content customers may be used as a basis of both recognizing and rewarding labour in the new economy (Adkins, 1995). Indeed, recent studies show how employees attempt in various ways to produce such customer effects vis-à-vis gender, and hence suggest that gender is increasingly an object of what, following Callon et al. (2002), can be termed a process of qualification–re-qualification. This takes expression in the ways managers, as well as employees, increasingly view gender as a fluid strategic artifice. For example, in recent studies of professional service work, workers describe gender as a self-conscious stratagem, that is as an artifice to be deployed in interactive service work (Gray, 2003).[10] Respondents participating in such studies describe their workplace performances of gender as strategically adaptable depending on the customer or client they will be dealing with. Indeed, as is the case with creative labour in regimes of cultural production organized by the brand, workers in the new economy attempt to make the signs of creative labour visible in their performances of gender in an attempt to ensure that this labour 'is seen and understood by the customer or the client' (Gray, 2003: 498), and hence attributed with value. Workers in the new economy therefore focus attention not on their labour as a form of property in the person – for instance on (the accumulation of) embodied skills and techniques as properties of the self – but on the effects of their labour (cultural work) on the intended audience. Indeed, rather than being owners of their labour as a form of property in the person, for workers in the new economy, rights to ownership of labour are adjudicated on the grounds of the effect of this labour in regard to the audience. In short, in the new economy, gender has been materially reconstituted from type or kind to cultural work and hence is subject to cycles of production, distribution and reception which make up contemporary regimes of cultural production.[11] This is witnessed in attempts to ensure a link between the labour process and audience effects, that is to insure a standardized creation of audience effects (and hence the creation of cultural value), via devices such as theme-ing and format franchising. Here workers are trained in the stylistics of performance and moreover a particular performance or customer interface is attributed to specific brands. Thus in the new economy, gender – along with other forms of cultural work – (and to use Callon's terms again) may be performatted in a variety of ways.

However, while this is the case, it is by no means certain that such customer effects can or will be made visible and recognized. Indeed, what gives further fuel to the idea that gender in the new economy is constituted as cultural work, is that there are various forms of contestation around both the visibility and recognition of such audience effects. This has been made particularly clear in regard to workplace performances of femininity by women (Adkins, 1995, 2001). Specifically, claims to ownership of this cultural work are sometimes denied to women since any customer effects generated by this labour are often defined (particularly by co-workers as well as customers) as the outcome of 'natural advantages' which should be neither recognized nor rewarded, for instance, by promotion. Any audience effects constituted by cultural labour are in other words rendered immanent (Adkins, 2002). What such forms of contestation in regard to the recognition of the effectivity of labour in the new economy suggest is that the links between audience effects and the labour of performance can in no way be taken for granted, indeed, that the political right to claim ownership of labour including the rights to claim workplace rewards is by no means straightforward. Specifically, there may be some significant exclusions from this right, and therefore from contemporary forms of (post-social contract) 'personhood'.

Conclusion

The claims in this article have been three-fold. First, that, in the new economy, gender has been materially reconstituted from type or kind to cultural product or work (which I have suggested is best understood as part of the broader historical process of type or kind becoming brand); second, that this has disentangled gender and the person with the implication that gender is now an alienable form of property subject to forms of qualification–re-qualification; and, third, that gender as a cultural work is not a form of property in the person but can only be laid claim to by less direct means and in particular via attempts to link audience effect with the labour process. I have suggested further that the example of gender is illustrative of a broad shift in the new economy whereby a social contract model of labour and of personhood – where actors are assumed to own, or at least to strive to accumulate property in the person – has been displaced by a model of labour and personhood whereby rights of ownership of creative works may only be claimed via the effects of the cultural product in regard to the intended audience.

I have attempted to illustrate this shift via a comparison of the relations between labour, ownership and people in the social contract and those in the new economy. Thus, while very few have achieved this ideal, under the conditions of the social contract, workers strove by various means to make claims to jurisdiction over property in the person, that is the political right to alienate or disentangle labour power. Welfarist regimes paradigmatic of the social contract, characterized by compacts between state, unions and employers, protected labour, and a strong gendered division of labour involving the idealized figures of the artisan and house-wife may be understood as arrangements which attempted to maximize the potential for claims to jurisdiction over labour power for some men. In the new economy, however, properties that in the social contract were assumed to reside in the person – embodied skills, techniques and expertise, which over time attach themselves to and were accumulated by the person, or perhaps more correctly were attached to the person via a variety of socio-technical and institutional arrangements such as apprenticeship and training systems – qualities which writers such as Pateman have referred to as abilities, bodily capacities, self-identity and self-esteem, are no longer organized in terms of relations of internality but are organized externally. That is, such properties are detached from, or perhaps more precisely cannot be assumed to be attached or to stick to the person in the new economy. As in the case of authorship, where there has been an historical shift away from the idea that authorship is a particular kind of property in the person, and is now laid claim to by other means (for instance, by making claims regarding the cultural effects of products), I have suggested that a similar shift is taking place in the case of labour in the new economy. Moreover, I have suggested that this reworking of the relations between people and their labour may be understood to be characteristic of (or in part what) defines the new economy. Thus, this article has aimed to contribute towards clarifying exactly what is new about the new economy, including how exactly it is that action in the new economy has taken on a more open and mediated character. I have argued further, that while not particularly interested in relations between labour and persons nonetheless, Callon's writings on economic activity provide useful tools for theorizing the reworking of material relations in new economy. This is so as Callon's rejection of the idea that classic notions of contract form the fundamental source or framing of economic action (such as social contract notions of property right) precisely opens out the question of the materiality of the new economy. Therefore, I have claimed that rather than Bourdieu (as is often suggested), it is Callon who provides important tools for theorizing material relations in the new economy since the former – via notions of property in the person (found, for instance, in the idea that various forms of capital may 'stick' to or be accumulated by the human subject) – always remains trapped within a social contract model of personhood.

As I have already mentioned, accounts of the new economy (including Callon's) have, however, been critiqued for dissolving and collapsing economic processes into culture, communications and information, and in particular for overlooking the fundamental framing of the economy – the relations of property and ownership which accompany the classic market exchange. Yet, as I hope my discussion of the shift of gender from type or kind to cultural work illustrates, this position may only be maintained if there is an acceptance of the view that culture is devoid of issues of ownership and property. Specifically, to accept this view is to miss out on the ways in which property and ownership are shifting in the new economy. This is to miss the ways in which ownership and property right are being restructured in such a way whereby it may never be entirely clear-cut, to paraphrase

Slater one last time, exactly when people are quits, since claims to ownership may be indirect, indeed, lie not in the hands of the 'owner' but in
attempts to link up audience effects with creative works and the labour
process. One further implication of the restructuring of ownership and
property right that I have outlined in this article is that the temporality of
exchange in the new economy cannot be taken for granted, but should be a
matter of investigation.

Finally, the restructuring of the relations between people and labour
which this article has underscored suggests that not only the temporality of
exchange needs to be reconsidered but also notions of personhood. As we
have seen, while Pateman has critiqued liberal notions of personhood by
arguing that property in the person can rarely be fully disentangled as
capacities, abilities and skills are internally (and not externally) related to
the person, nonetheless her account still accepts the (normative) view that
personhood may be achieved via claims regarding the ownership of property
in the person. Thus, for Pateman, while many (notably women) have been
historically excluded from this right, certain men have acquired jurisdiction over property in the person and hence achieved – or have come close
to achieving – the modern status of the individual. Yet in the new economy
organized by cultural principles of the brand, sovereignty over property in
the person – including claims that capacities, skills and abilities stick to or
can be accumulated by the person – does not constitute the ground for
ownership of labour. I have suggested that in this context a social contract
model of personhood cannot and does not form a basis for understanding
the organization of labour and the relations between labour and persons.
The analysis of the relations between people and labour offered in this
article suggests that in new economy we may be witnessing the end of
property in the person, that is, the end of modern notions of personhood.

Notes

1. Callon's analysis of economic activity has been accused of ahistoricism.
However, my claim here is that it is particularly relevant for the new economy,
indeed, that Callon may be regarded as a theorist of the new economy.

2. As is well known, a central part of Pateman's argument was that the social
contract is the means through which modern patriarchy is constituted, a form of
rule which, she argued, is defined not via paternity but by fraternity (or what some
theorists have referred to as public patriarchy).

3. Lury's account, for example, is more concerned with shifts in the organization
of cultural value than with the organization of labour (albeit that the two are related).

4. This may be seen as part of the de-differentiation of social and cultural fields
(Lash, 1990), and more generally may be understood as illustrative of how the
economy is playing an increasingly important role in the production and reproduction of culture, or what is sometimes referred to as a shift from the political to
the cultural economy. However, while many authors understand this shift as involving an increasing significance of signs (that commercial objects have become more
sign-like) together with a decreasing significance of materiality, here, and in

agreement with Slater (2002b), I am suggesting that materiality in the new economy is being reworked in some significant ways.

5. These techniques may be understood as examples of the modes of power operative within Deleuze's control society (see Brown, 2003).

6. While the focus of this article is the shifting relations between labour and people and the emergence of gender as a cultural work or product as being illustrative of this shift, nonetheless the analysis proposed here also has important implications for understandings of gender. Specifically, it suggests that in post-social or network society, object rather than subject relations may be increasingly important in understanding the organization of gender. This is a significant point as much modern gender theory posits gender as a mode of subjectivity, that is, as a form of interiority.

7. This is often expressed via Goffman's understanding of performance to define the performance of gender in the new economy.

8. While these points have been noted in such studies, the pervasive assumption of a social contract model of personhood has disallowed a consideration of their full implications.

9. The notion of interactant has been used by Wittel et al. (2002) to describe the ways in which 'representation' in the new media differ from representations in older forms of cultural production. Specifically, it is used to describe how in new media the distance between, for example, a novel and a reader or a film and spectator has collapsed, as the spectator is now a player, or put differently, the subject is now in the (media) object. While this has been argued specifically for new media objects, the analysis presented in this article suggests that the collapse of the gap between production and audiences is also an issue for labour in the new economy.

10. For many commentators on the new economy, this strategic deployment of gender by workers is understood to undermine the ways in which gender was previously arranged so as to produce disadvantage, particularly for women at work, indeed, even to transgress gender. Thus, performative arrangements of gender in the new economy are often celebrated. However, while it may be the case that that the rearrangement of gender from type or kind to cultural product undoes socio-structural arrangements of gender, it should not necessarily be assumed that new forms of domination do not arise in this context.

11. This is not an argument simply to suggest that the socialized or constructed character of gender is somehow revealed through its expression as cultural labour in the new economy, but rather that the broad historical process of type or kind becoming brand has radically reworked the relations between labour power as property and gender. So while in industrial society struggles over the right to disentangle labour power in significant part defined gender (i.e. the social contract was a sexual contract), in network society or post-social relations, gender becomes alienable as (cultural) property. Put differently, this article underscores a shift not only in the relations between people and their labour, but also a significant and important shift between gender, labour power and the labour process.

References

Adkins, Lisa (1995) *Gendered Work: Sexuality, Family and the Labour Market.* Buckingham: Open University Press.

Adkins, Lisa (2000) 'Objects of Innovation', in S. Ahmed, J. Kilby, C. Lury, M.

McNeil and B. Skeggs (eds) *Transformations: Thinking Through Feminism*. London: Routledge.

Adkins, Lisa (2001) 'Cultural Feminization: "Money, Sex and Power" for Women', *Signs: Journal of Women in Culture and Society* 26(3): 31–57.

Adkins, Lisa (2002) *Revisions: Gender and Sexuality in Late Modernity*. Buckingham: Open University Press.

Adkins, Lisa and Celia Lury (1999) 'The Labour of Identities: Performing Identities, Performing Economies', *Economy and Society* 28(4): 598–614.

Beck, Ulrich, Anthony Giddens and Scott Lash (1994) *Reflexive Modernization: Politics, Tradition and Aesthetics in the Modern Social Order*. Cambridge: Polity.

Benjamin, Walter (1970) 'The Work of Art in the Age of Mechanical Reproduction', in *Illuminations*. London: Collins.

Beynon, Huw, Damian Grimshaw, Jill Rubery and Kevin Ward (2002) *Managing Employment Change: The New Realities of Work*. Oxford: Oxford University Press.

Brown, Megan (2003) 'Survival at Work: Flexibility and Adaptability in American Corporate Culture', *Cultural Studies* 17(5): 713–33.

Callon, Michel (1998a) 'Introduction: The Embeddedness of Economic Markets in Economics', in M. Callon (ed.) *The Laws of the Market*. Oxford: Blackwells/Sociological Review.

Callon, Michel (1998b) 'An Essay on Framing and Overflowing', in M. Callon (ed.) *The Laws of the Market*. Oxford: Blackwells/Sociological Review.

Callon, Michel, Cecile Meadel and Vololona Rabehariosa (2002) 'The Economy of Qualities', *Economy and Society* 31(2): 194–217.

Carrier, James and Daniel Miller (eds) (1998) *Virtualism: A New Political Economy*. Oxford: Berg.

Casey, Catherine (1995) *Work, Self and Society: After Industrialism*. London: Routledge.

Castells, Manuel (1996) *The Rise of the Network Society*. Oxford: Blackwell.

Castells, Manuel (2000) 'The Institutions of the New Economy', plenary address, *Delivering the Virtual Promise?* Available online at: http://virtualsociety.sbs.ox.ac.uk/texts/events/castells.htm

Du Gay, Paul (1996) *Consumption and Identity at Work*. London: Sage.

Frow, John (2002) 'Signature and Brand', in J. Collins (ed.) *High-Pop: Making Culture into Popular Entertainment*. Oxford: Blackwell.

Gray, Ann (2003) 'Enterprising Femininity: New Modes of Work and Subjectivity', *European Journal of Cultural Studies* 6(4): 489–506.

Halford, Susan, Mike Savage and Anne Witz (1997) *Gender, Careers and Organizations: Current Developments in Banking, Nursing and Local Government*. Basingstoke: Macmillan.

Haraway, Donna J. (1997) *Modest_Witness@Second_Millennium. FemaleMan©_Meets_OncoMouse™: Feminism and Technoscience*. London: Routledge.

Hardt, Michael and Antonio Negri (2000) *Empire*. Cambridge, MA: Harvard University Press.

Hinchliffe, Steve (2000) 'Performance and Experimental Knowledge: Outdoor Training and the End of Epistemology', *Society and Space* 18: 575–95.

Hochschild, Arlie R. (1983) *The Managed Heart: The Commercialization of Human Feeling*. Berkeley: University of California Press.

Knorr-Cetina, Karin (2004) 'Capturing Markets? A Review Essay on Harrison White on Producer Markets', *Socio-Economic Review* 2: 137–47.

Knorr-Cetina, Karin and Urs Bruegger (2000) 'The Market as an Object of Attachment: Exploring Postsocial Relations in Financial Markets', *Canadian Journal of Sociology* 25(2): 141–68.

Knorr-Cetina, Karin and Urs Bruegger (2002) 'Traders' Engagement with Markets: A Postsocial Relationship', *Theory, Culture & Society* 19(5/6): 161–85.

Lash, Scott (1990) *Sociology of Postmodernism*. New York: Routledge.

Lash, Scott (1994) 'Reflexivity and its Doubles: Structure, Aesthetics, Community', in U. Beck, A. Giddens and S. Lash (eds) *Reflexive Modernization: Politics, Traditions and Aesthetics in the Modern Social Order*. Cambridge: Polity Press.

Lash, Scott (2002) *Critique of Information*. London: Sage.

Lovell, Terry (2000) 'Thinking Feminism With and Against Bourdieu', *Feminist Theory* 1(1): 11–32.

Lury, Celia (1993) *Cultural Rights: Technology, Legality and Personality*. New York: Routledge.

Lury, Celia (2000) 'The United Colors of Diversity: Essential and Inessential Culture', in S. Franklin, C. Lury and J. Stacey *Global Nature, Global Culture*. London: Sage.

Lury, Celia (2003) 'The Game of Loyalt(o)y: Diversions and Divisions in Network Society', *Sociological Review* 51(3): 301–20.

Martin, Emily (1994) *Flexible Bodies: Tracking Immunity on American Culture: From the Days of Polio to the Age of AIDS*. Boston, MA: Beacon.

Martin, Emily (2000) 'Flexible Survivors', *Cultural Values* 4(4): 512–17.

Pateman, Carole (1988) *The Sexual Contract*. Cambridge: Polity Press.

Pateman, Carole (2002) 'Interview with Carole Pateman: The Sexual Contract, Women in Politics, Globalization and Citizenship', *Feminist Review* 70: 123–33.

Peters, Tom (1999) *The Brand You*. New York: Random House.

Rodowick, D.N. (2001) *Reading the Figural, or, Philosophy after the New Media*. Durham, NC: Duke University Press.

Sennett, Richard (1998) *The Corrosion of Character: The Personal Consequences of Work in the New Capitalism*. New York: Norton.

Slater, Don (2002a) 'From Calculation to Alienation: Disentangling Economic Abstractions', *Economy and Society* 31(2): 234–49.

Slater, Don (2002b) 'Markets, Materiality and the "New Economy"', in S. Metcalfe and A. Warde (eds) *Market Relations and the Competitive Process*. Manchester: Manchester University Press.

Thrift, Nigel (1998) 'Virtual Capitalism: The Globalization of Reflexive Business Knowledge', in J. Carrier and D. Miller (eds) *Virtualism: A New Political Economy*. Oxford: Berg.

Wittel, Andreas (2001) 'Towards a Network Sociality', *Theory, Culture & Society* 18(6): 51–76.

Wittel, Andreas, Celia Lury and Scott Lash (2002) 'Real and Virtual Connectivity: New Media in London', in S. Woolgar (ed.) *Virtual Society: Technology, Cyberbole, Reality*. Oxford: Oxford University Press.

Lisa Adkins is Reader in Sociology at the University of Manchester. She has published widely in the fields of the sociology of post-industrial economies, feminist social theory, and the sociology of gender and sexuality. Her latest book is *Revisions: Gender and Sexuality in Late Modernity* (Open University Press, 2002).

Computing the Human

N. Katherine Hayles

A MONG THE intriguing unwritten books are those exploring the influence of the future on the present. Who wouldn't leap at the chance to review the non-existent *Influences of the Twenty-First Century on the Nineteenth?* As this imaginary book would undoubtedly testify, visions of the future, especially in technologically advanced eras, can dramatically affect present developments. Of special interest, then, is the spate of recent works projecting a future in which humans and intelligent machines become virtually indistinguishable from one another. Through such emerging technologies as neural implants, quantum computing, and nanotechnology, humans will become computationally enhanced and computers will become humanly responsive until in a mere 100 years, by Ray Kurzweil's reckoning, we can expect that both humans and computers will be so transformed as to be unrecognizable by present standards (1999: 280). Migrating their minds from one physical medium to another as convenience dictates, these future entities will become effectively immortal, manifesting themselves in forms that are impossible to categorize as either humans or machines.[1]

As Kurzweil himself acknowledges, however, nothing is more problematic than predicting the future. If the record of past predictions is any guide, the one thing we can know for sure is that when the future arrives, it will be different from the future we expected. Instructed by the pandemic failure to project accurately very far into the future, my interest here is not to engage in this kind of speculation but rather to explore the influence that such predictions have on our *present* concepts.[2] At stake, I will argue, is not so much the risky game of long-term predictions as contending for how we now understand human thinking, acting, and sensing – in short, how we understand what it means to be human.[3]

The complex interactions shaping our ideas of 'human nature' include material culture. Anthropologists have long recognized that the construction of artifacts and living spaces materially affects human evolution. Changes in the human skeleton that permitted upright walking co-evolved, anthropologists believe, with the ability to transport objects, which in turn led to the development of technology. We need not refer to something as contemporary and exotic as genetic engineering to realize that for millennia, a two-cycle phenomenon has been at work: humans create objects, which in turn help to shape humans. This ancient evolutionary process has taken a new turn with the invention of intelligent machines. As Sherry Turkle (1984) has demonstrated in her study of how children interact with intelligent toys, artifacts that seem to manifest human characteristics act as mirrors or 'second selves' through which we re-define our image of ourselves. The simulations, software, and robots we have today fall far short of human accomplishments (though in other ways they exceed what humans can do, for example, in detecting subtle patterns in large data sets). Nevertheless, researchers with the greatest stake in developing these objects consistently use a rhetoric that first takes human behavior as the inspiration for machine design, and then, in a reverse feedback loop, reinterprets human behavior in light of the machines.

To illustrate this process and explore its implications, I will focus on what is known in the field of artificial intelligence as the Sense–Think–Act paradigm (STA). It is no mystery why researchers concentrate on STA, for it defines the necessary behaviors an entity needs to interact with the world. Sensing allows the entity to perceive the world, while cognition processes the sensory data and prepares for the next step, action.

At each node of the STA paradigm we will see similar dynamics at work, although there are also important differences that distinguish between research programs. Constant across all three nodes, however, is a tendency to extrapolate from relatively simple mechanical behaviors to much more complex human situations and a consequent redescription of the human in terms of the intelligent machine. These machines constitute, we are told, a new evolutionary phylum that will occupy the same niche as *Homo sapiens*, an equivalence implied in Peter Menzel and Faith D'Aluisio's (2000) name for this species, *Robo sapiens*. The pressure to see *Homo sapiens* and *Robo sapiens* as essentially the same emerges as a narrative of progress that sees this convergence as the endpoint of human evolution.

Whether or not the predicted future occurs as it has been envisioned, the effect is to shape how human being is understood *in the present*. Those who want to argue for the uniqueness of human nature, like Francis Fukuyama (2002), are forced (consciously or unconsciously) to concentrate on those aspects of human behaviors that machines are least likely to share. Others who envision a convergence between humans and robots, like Moravec (1990, 1999) and Kurzweil (1999), de-emphasize those aspects of human nature that intelligent machines do not share, such as embodiment. Whether one resists or accepts the convergence scenario, the relation between humans and intelligent machines thus acts as a strange attractor, defining the phase space within which narrative pathways may be traced. What becomes difficult to imagine is a description of the human that does not take the intelligent machine as a reference point. This perspective is arguably becoming the dominant framework in which highly developed countries such as the USA understand the future. Whatever the future, the implications of this perspective for the present are consequential. Later I will return to these questions to evaluate the various arguments and positions. First, however, it will be useful to explore the basis for the convergence scenario in research that has emerged around the STA paradigm.

Acting

Among the influential researchers defining our present relation to intelligent machines is Rodney Brooks. Brooks (2002) describes the oppositional heuristic at the center of his research method. He looks for an assumption that is not even discussed in the research community because it is considered to be so well established; he then supposes that this 'self-evident truth' is not true. When he first began his research, researchers assumed that artificial intelligence should be modeled on conscious human thought. A robot moving across a room, for example, should have available a representation of the room and the means to calculate each move to map it onto the representation. Brooks believed this top-down approach was much too limiting. He saw the approach in action with a room-crossing robot designed by his friend and fellow student, Hans Moravec. The robot required heavy computational power and a strategy that took hours to implement, for each time it made a move, it would stop, figure out where it was, and then calculate the next move. Meanwhile, if anyone entered the room it was in the process of navigating, it would be hopelessly thrown off and forced to begin again. Brooks figured that a cockroach could not possibly have as much computational power on board as the robot, yet it could accomplish the same task in a fraction of the time. The problem, as Brooks saw it, was the assumption that a robot had to operate from a representation of the world.

Brooks' oppositional strategy, by contrast, was to build from the bottom up rather the top down. One of his inspirations (2002: 17–21) was William Grey Walter, who in the 1940s built small robots, dubbed electric tortoises, that could robustly navigate spaces and return to the hutches for refueling when their batteries ran low. Following this lead, among others, Brooks began to design robots that could move robustly in the world without any central representation; he is fond of saying these robots take 'the world as its own best model'. Incorporating a design principle he calls 'subsumption architecture', the robots were created using a hierarchical structure in which higher level layers could subsume the role of lower levels when they wanted to take control. In the absence of this control, the lower layers continued to carry out their programming without the necessity to have each move planned from above. Each layer was constituted as a simple finite state machine programmed for a specific behavior with very limited memory, often less than a kilobyte of RAM. The semi-autonomous layers carried out their programming more or less independently of the others. The architecture was robust, because if any one level failed to work as planned, the other layers could continue to operate. There was no central unit that would correspond to a conscious brain, only a small module that adjudicated conflicts when the commands of different layers interfered with each other. Nor was there any central representation; each layer 'saw' the world differently with no need to reconcile its vision of what was happening with the other layers.

An example of a robot constructed using this model is Genghis, a six-legged insectile robot 30 cms long. The robot is programmed to prowl and engage in exploratory behaviors; when it senses a human in its vicinity, it charges toward him/her, albeit at such a slow pace that the human is in no danger of being overtaken. Instead of programming the gait in detail – a fearsomely complex computational challenge – each leg is coordinated only very slightly with the others, and the gait emerges from the local semi-autonomous behavior of each leg following its programming independently. Embedded in the design is a concept central to much of Brooks' research:

> Complex (and useful) behavior need not necessarily be a product of an extremely complex control system. Rather, complex behavior may simply be a reflection of a complex environment . . . It may be an observer who ascribes complexity to an organism – not necessarily its designer. (1999: 7)

A variation on this idea is the 'cheap trick', a behavior that emerges spontaneously from the interaction of other programmed behaviors. When detractors pointed out that these robots merely operated at an insectile level of intelligence, Brooks responded by pointing out that on an evolutionary timeline, the appearance of insects occurred at 90 percent of the time it took to evolve humans. This suggests, Brooks contends, that the hard problem is robust movement in an unpredictable complex three-dimensional environment. Once this problem is solved, higher level cognitive functioning can be evolved relatively easily.

To illustrate the idea of taking the world as its own best model, Brooks and Maya Mataric, then a graduate student in the Artificial Intelligence Laboratory at MIT, built Toto, a path-following robot described in Brooks (1999: 37–56). Toto represented an advance over Genghis because it was able to operate as if it had dynamically changing long-term goals and maps it had built up over time. It accomplished these goals, however, without any central representation of the spaces it navigated. Rather, it detected landmarks and then stored these locations in nodes connecting to one another in a spreading activation topology. Each node was capable of comparing the incoming data about landmarks to its own position; navigation then became a matter of following the shortest path through the locations stored in the nodes to arrive at the desired goal. This clever arrangement in effect folded the map-making activity back onto the controller, so that the 'map' existed not as an abstract representation but rather emerged dynamically out of the robot's exploratory behavior. Commenting from his perspective a decade later on the article he and Mataric co-published on this work, Brooks writes:

> This paper is very confusing to many people. It turns the whole notion of representation upside down. The robot never has things circumscribed in its head, but it acts just like we imagine it ought to. I view this work as the nail in the coffin of traditional representation. (1999: 37)

Extrapolated to humans (as Brooks is not slow to do), these results implied that complex behaviors can emerge out of simple operations that do not depend on a central representation of the world. Students working with Brooks took to accompanying people home from work to determine how much of their navigation was conscious and how much was on 'automatic pilot'. Consciousness quickly became relegated, in the term used by many researchers in artificial life and intelligence, to the role of an 'epiphenomenon'. A late evolutionary development, consciousness was seen to play a much smaller role in action than had traditionally been thought. Experiments by Benjamin Libet, for example, showed that when a human subject was asked to indicate when he had decided to raise his arm, muscles were already set in action before he spoke, suggesting that the decision-making process was arriving at consciousness *after* the decision had already been made (Libet, 1985; Haggard and Libet, 2001).[4] Moreover, sensory data were known to arrive at consciousness already highly processed, so that much of the interpretation of sensory data had already been made before the conscious mind was aware of it. Brooks cites as evidence that the conscious mind was operating from a very partial view of the world – the fact that we go through life with a large blank spot in the middle of our visual field, although we are not aware of it. At the same time, the neurophysiologist Antonio Damasio (1995, 2000) was arguing from data gathered from thousands of patients that much cognition goes on in lower brain regions such as the limbic system, the peripheral nervous system, and the viscera. A joke told by the comedian Emo Phillips is relevant here. 'I used to think the brain was the most wonderful organ in the body,' he says. 'Then I asked myself, "Who's telling me this?" '

These results led Brooks to explore how far the experimental program he had implemented with insect-like robots could be carried out with a higher-level robot meant to evolve more sophisticated human behaviors. This agenda led to Cog (a play on cognition, as well as a mechanical cog), a head-and-torso robot with human-like eye motions and software that enables it to interact on a limited basis with human partners. Its eyes can saccade, smoothly follow, and fixate on objects in the room, using actively moving cameras. It can detect faces and recognize a few of them, as long as it has a full frontal view. It can pick out saturated colors and recognize skinlike colors across the range of human skin tones. It also has the possibility to create emergent behaviors more complex than those that were programmed in. This potential became clear when one of the graduate students who helped designed it, Cynthia Breazeal, held up a whiteboard eraser and shook it to get Cog's attention. Cog then reached and touched it, whereupon the game repeated. Watching the videotape of these interactions, Brooks reports that

> it appeared that Cog and Cynthia were taking turns. But on our development chart we were years away from programming the ability to take turns into Cog. The reality was that Cynthia was doing all the turn-taking, but to an external observer the source of the causation was not obvious. (2002: 91)

Drawing on this insight, Breazeal decided to build for her PhD project a robot that could engage in social interactions. The result was Kismet, whose features are specifically designed to encourage emotional responses from humans. Kismet has moveable eyebrows, oversized eyeballs that are human-like in appearance with foveal cameras behind them, and actuators that let it move its neck across three different axes. It is controlled by 15 different computers, some moving parts of the face and eyes and others receiving visual and audio inputs. Following the idea of semi-autonomous layers, Kismet has no central control unit but rather a series of distributed smaller units. Its software drives it to seek human interaction, with a set of internal drives that increase over time until they are satisfied. These include searching for saturated colors and skin tones, which causes the robot to fixate on toys and people. It appears as if the robot is searching for play objects, but as Brooks (2002: 94) observes, 'the overall behavior emerges from the interactions of simpler behaviors, mediated through the world.' It displays expressions typical of emotional responses, and it can put prosody into its voice as well as discern prosody in human voices. Its software is programmed to follow the basic mechanism of turn-taking in conversation, although Kismet cannot actually understand what is said to it nor say anything meaningful in return. When naive observers (i.e., those who did not know about the robot's programming) were brought into the lab and asked to talk to Kismet, most of them were able to engage in 'conversation' with Kismet, although the robot babbled only nonsense syllables, much as a human infant might. An important part of this project is a perspective that considers the robot not as an isolated unit but rather as part of an ecological whole that includes the humans with which it interacts. Human interpretation and response make the robot's actions more meaningful than they otherwise would be. The community, understood as the robot plus its human interlocutors, is greater than the sum of its parts because the robot's design and programming have been created to optimize interactions with humans.

These projects, focusing on the dynamics of embodied robots, have made Brooks more cautious than researchers like Kurzweil and Moravec about the possibilities for creating artificial bodies that human consciousness can inhabit. Discussing their versions of techno-utopianism, Brooks comments that although this 'strong version of salvation seems plausible in principle',

> we may yet be hundreds of years off in figuring out just how to do it. It takes computational chauvinism to new heights. It neglects the primary role played by the bath of neurotransmitters and hormones in which our neuronal cells swim. It neglects the role of our body in placing constraints and providing

noncomputational aspects to our existence. And it may be completely missing the *juice*. (2002: 206)

The 'juice' is Brooks' term for as-yet-undiscovered aspects of human cognition, evolution and biology that would provide new avenues for research in artificial intelligence. The contrarian methodology he follows leads in this instance to the idea that there may be lying in wait some entirely new principles that will revolutionize artificial intelligence, a hope undergirding the 'Living Machines' projects on which his graduate students are currently engaged. Here the influence of the future can be seen not in long-range predictions but rather in a shotgun methodology in which a wide variety of approaches are tried in the hope that one or more may pay off. Before assessing these possibilities and their implications for our present understanding of what it means to be human, I turn now to another node of the STA paradigm.

Sensing

Sensors, essential to robust movement in the world and crucial to the development of free-ranging robots, are rapidly developing along many sensory modalities, including visual, auditory, tactile and infrared. In the interest of tracking how visions of the future are affecting our present vision of the human, I will leave aside these mainstream developments to consider a rather quirky and quixotic proposal to develop epistemically autonomous devices, first advanced by Peter Cariani. Cariani (1991) arrived at this idea by critiquing existing models of artificial life, especially the idea of emergence. He noted that emergence risks becoming entangled in Descartes' Dictum, which states that if our devices follow our specifications exactly, they will never go beyond them, remaining mired in the realm of classical mechanics where devices perform exactly as predicted with nothing new added. On the other hand, if the devices depart too freely from our specifications, they are unlikely to be useful for our purposes. To clarify how emergence can generate novelty and still remain useful, Cariani defined emergence relative to a model. If the processes that link symbols to the world (for example, processes that link ones and zeros to changing voltages in a computer) result in new functions, then the system has extended the realm of its symbolic activity. For example, a third value might emerge that is neither one nor zero. If the innovation takes the form of new content for the symbols, it is said to be semantically emergent; if it arranges the symbols in a new way, it is syntactically emergent. Either of these conditions leads to new observational primitives. This point is important, Cariani argues, because systems that can only consider primitives specified by their designers remain constrained by the assumptions implicit in those specifications. In this sense the system can know the world only through the modalities dictated by its designer. Although it might work on these data to create new results, the scope of novelty is limited by having its theater of operations – the data that create and circumscribe its world – determined in advance without the possibility of free innovation. For maximum novelty, one needs a system that can break out of the frame created by the designer, deciding what will count as inputs for its operations. Such a system can then become epistemically autonomous relative to its creator, 'capable of searching realms for which we have no inkling' (Cariani, 1991: 779).

One way to achieve epistemic autonomy is to have sensors constructed by the system itself instead of being specified by the designer. Searching the literature for examples, Cariani found only one, a device created in the 1950s by the cybernetician Gordon Pask, who demonstrated it at various conferences under the name 'Pask's Ear'. The system was a simple electrochemical device consisting of a set of platinum electrodes in an aqueous ferrous sulfate/sulfuric acid solution. When current is fed through the electrodes, iron threads tend to grow between the electrodes. If no current passes through a thread, it dissolves back into the acidic solution. Branches form off the main threads, setting up a situation in which various threads compete for the available current. Generally the threads that follow the path of maximum current flourish the best, but the dynamic becomes complicated when threads join and form larger collaborative structures. In the complex growth and decay of threads, the system mimics an evolutionary ecology that responds to rewards (more current) and punishment (less current). More current does not specify the form that the growth will take; it only establishes the potential for more growth. The system itself discovers the optimum form for the conditions, which turn out to include other factors in the room environment, such as temperature, magnetic fields, and vibrations from auditory signals. Capitalizing on the fact that the system was capable of taking auditory signals as input, Pask 'trained' the system to recognize different sound frequencies by sending through more current at one frequency than another. Within half a day, he was able to train the system to discriminate between 50Hz and 100Hz tones. Such a system, Cariani (1991: 789) argues, 'would be epistemically autonomous, capable of choosing its own semantic categories as well as its syntactic operations on the alternatives'. Cariani (1998: 721) imagines that similar methodologies might be used to create new signaling possibilities in biological neurons.

Building on Cariani's ideas, Jon Bird and Paul Layzell (2002) built an 'evolved radio'. They state clearly the motivations for this research. Reasoning along lines familiar from Rodney Brooks, Bird and Layzell consider the constraints on simulations in contrast to real-world modeling, now with an emphasis on sensing rather than acting:

> [There is] a fundamental constraint in simulating sensor evolution: the experimenter sets a *bound* on the possible interactions between the agent and the environment. This is a direct consequence of the simulation process: firstly, the experimenter has to model *explicitly* how different environmental stimuli change the state of the sensors; secondly, experimenters only simulate those aspects of the environment that they think are relevant to their experiment, otherwise the simulation would become computationally intractable.

These constraints make it very difficult to see how there can be a simulation of novel sensors.

They continue:

> Novel sensors are constructed when a device, rather than an experimenter, determines which of the infinite number of environmental perturbations act as useful stimuli. (2002: 2)

Working from this perspective, they noticed reports in the literature of small inductance and capacitance differences emerging spontaneously among transistor circuits in a new form of evolvable hardware called a Field Programmable Gate Array. They thought they could opportunistically capitalize on this emergent property by building an 'evolvable motherboard' using a matrix of analogue switches, which are themselves semiconductor devices.

Radio circuits are comprised of oscillators created when resistors are used to control the charge release of capacitors, according to the well-known RC time constant. Bird and Layzell wanted to arrange the transistors in appropriate patterns so that these oscillators would emerge spontaneously. To 'kick-start the evolutionary process', Bird and Layzell (2002: 2) rewarded frequency, amplitude of oscillation, and output amplitude. Once they had succeeded in creating the desired oscillators, the oscillators acted as a radio by picking up on the waves generated by the clocks of nearby PC computers. To say the emergent circuits were quirky would be to indulge in understatement. Some would work only when a soldering iron on a nearby workbench was plugged in, although it did not have to be on. Other circuits would work only when the oscilloscope was on. In effect, the evolved radio took the entire room as its environment, using the room's resources in ways that were not determined by the researchers and probably not fully known by them. Citing Richard Lewontin, Bird and Layzell point out that the environment can be theoretically partitioned into an infinite number of niches, but it takes an organism exploiting a niche for it to be recognized as such. Some organisms, they point out, have adapted in highly specialized ways to niches that remain relatively constant; general solutions are usually found only by organisms that inhabit highly variable evolutionary niches. The evolved radio is like a highly specialized organism, exploiting the specific characteristics of the room and unable to adapt if the room's configurations change. This disadvantage notwithstanding, the advantage is that the system itself establishes the nature of its relation to the world. It decides what it will recognize as relevant inputs and in this sense evolves its own sensors.

Evolving new sensors implies constructing new worlds. As Cariani observes:

> . . . sensors determine the perceptual categories that are available, while effectors determine the kinds of primitive actions that can be realized. Sensors and

effectors thus determine the nature of the external semantics of the internal, informational states of organisms and robotic devices. (1998: 718)

While humans have for millennia used what Cariani calls 'active sensing' – 'poking, pushing, bending' – to extend their sensory range and for hundreds of years have used prostheses to create new sensory experiences (for example, microscopes and telescopes), only recently has it been possible to construct evolving sensors and what Cariani (1998: 718) calls 'internalized sensing', that is, '"bringing the world into the device" by creating internal, analog representations of the world out of which internal sensors extract newly-relevant properties'.

We can draw several connections between Cariani's call for research in the frontiers of sensor research and the future of the human. One implication, explicitly noted by Jon Bird and Andy Webster (2001), is the blurring of the boundary between creator and created; humans create autonomous systems in the sense that they set them running, but a large measure of the creativity in parsing the world is created by the system itself. Other implications emerge from the physical and informational integration of human sensory systems and artificial intelligence. Kevin Warwick's recent implant chip with a 100 electrode array into the median nerve fibers of his forearm is an example of 'bringing the world into the device' by connecting the human nervous system with new internal sensors. Warwick's implant communicates both with the external world and his own nervous system. Although it is not clear yet how these neural connections might affect his perceptions – if at all – the import is clear. Perceptual processing will increasingly be mediated through intelligent components that feed directly into the human nervous system, much as William Gibson imagined in the cyberspace decks of the *Neuromancer* trilogy. More mundane examples are increasingly evident, for example, the night vision goggles worn by US troops in the 1991 Gulf War. When the human nervous system is receiving information through prostheses seamlessly integrated with internal implants, the line between human sensing and the sensing capabilities of intelligent machines becomes increasingly blurred. 'Machines R Us' is one interpretation of the permeable boundary between the sensing 'native' to humans and the sensing done through networks of intelligent software and hardware that communicate, directly and indirectly, with the human nervous system.

Another conclusion emerges from Cariani's call (1998) for research in sensors that can adapt and evolve independently of the epistemic categories of the humans who create them. The well-known and perhaps apocryphal story of the neural net trained to recognize army tanks will illustrate the point. For obvious reasons, the army wanted to develop an intelligent machine that could discriminate between real and pretend tanks. A neural net was constructed and trained using two sets of data, one consisting of photographs showing plywood cutouts of tanks and the other actual tanks. After some training, the net was able to discriminate flawlessly between the situations. As is customary, the net was then tested against a third data set showing pretend and real tanks in the same landscape; it failed miserably. Further investigation revealed that the original two data sets had been filmed on different days. One of the days was overcast with lots of clouds, and the other day was clear. The net, it turned out, was discriminating between the presence and absence of clouds. The anecdote shows the ambiguous potential of epistemically autonomous devices for categorizing the world in entirely different ways from the humans with whom they interact. While this autonomy might be used to enrich the human perception of the world by revealing novel kinds of constructions, it also can create a breed of autonomous devices that parse the world in radically different ways from their human trainers.

A counter-narrative, also perhaps apocryphal, emerged from the 1991 Gulf War. US soldiers firing at tanks had been trained on simulators that imaged flames shooting out from the tank to indicate a kill. When army investigators examined Iraqi tanks that were defeated in battles, they found that for some tanks the soldiers had fired four to five times the amount of munitions necessary to disable the tanks. They hypothesized that the overuse of firepower happened because no flames shot out, so the soldiers continued firing. If the hypothesis is correct, human perceptions were altered in accord with the idiosyncrasies of intelligent machines, providing an example of what can happen when human–machine perceptions are caught in a feedback loop with one another. Of course, humans constantly engage in perceptual feedback loops with one another, a phenomenon well known for increasing the stability of groups sharing the consensual hallucination. By contrast, in Greg Bear's *Darwin's Radio* (2004) children are born with genetic mutations caused by the reactivation of ancient retroviruses – their development of entirely new sensors is one of the indications that they have become a new species vastly superior to the *Homo sapiens* they will supersede. The realization that novel sensors may open new evolutionary pathways for both humans and intelligent machines is one of the potent pressures on our current conceptions of what it means to be human.

Thinking

John Koza was tired. He had heard his scientific colleagues complain too many times that artificial life, while conceptually interesting, was only capable of solving toy problems that were not much use in the real world. Koza, one of the pioneers in developing genetic programming, specializes in the creation of software that can evolve through many generations and find new solutions, not explicitly specified by the designer, to complex problems. Inspired by biological evolution, the basic idea is to generate several variations of programs, test their performance against some fitness criteria, and use the most successful performers as the genetic 'parents' of the next generation, which again consists of variations that are tested in turn, and so on until the solutions that the programs generate are judged to be successful. Koza (Koza et al., 1999: 5) was particularly interested in creating programs that, as he puts it, could arrive at solutions 'competitive with human-produced results'.

We might call this the Koza Turing test, for it introduces consequential alterations that expand the range and significance of Turing's classic test. Recall that Turing proposed to settle the question of whether computers could think by asking a human interlocutor to question a human and a computer, respectively. If the interlocutor was unable to distinguish successfully between the two on the basis of answers they submitted to his questions, this constituted *prima facie* evidence, Turing argued, that machines could think. By operationalizing the question of intelligence, Turing made it possible to construct situations in which the proposition that machines can think could be either proved or disproved, thus removing it from the realm of philosophical speculation to (putative) empirical testing. Once this move has been made, the outcome is all but certain, for researchers will simply focus on creating programs that can satisfy this criterion until they succeed. The proof of the pudding lies less in the program design than in arriving at a consensus for the test. Like a magician that distracts the audience's attention by having them focus on actions that occur *after* the crucial move has already been made, the Turing test, through its very existence, already presupposes consensus on the criteria that render inevitable the conclusion that machines can think.

Reams of commentaries have been written on the subtleties of the Turing test and its implications for human–computer interactions. Evaluating this substantial body of scholarship is beyond the scope of this article, but suffice it to say that Koza's emphasis on producing human-competitive results significantly shifts the focus. At stake is not whether machines are intelligent – a question I consider to be largely answered in the affirmative, given the cognitively sophisticated acts contemporary computer programs can perform – but whether computers can solve problems that have traditionally been regarded as requiring intuitive knowledge and creativity. Like Turing, Koza proposes to operationalize the question of creativity, thereby rendering it capable of proof or disproof. Among other criteria, he proposes that the programs should be judged as producing human-competitive results if they generate results that have received a patent in the past, improve on existing patents, or qualify as a patentable new invention in the present. Alternatively, they should also be judged as human-competitive if the result is equal or better than a result accepted as scientifically significant by a (human) peer-reviewed journal.

To tackle this challenge, Koza and his co-authors (1999) created genetic programs that could design bandpass filters, that is, electrical circuits capable of distinguishing between and separating out signals of one frequency versus another. There is no explicit procedure for designing these filters because it is desirable to optimize a number of different criteria, including the sharpness of separation, parsimony of components, and so forth. Electrical engineers who specialize in these designs rely on a large amount of intuitive knowledge gained through years of experience. Koza's algorithm worked by starting with extremely simple circuits – kindergarten level. The program then created different variations, tested them, chose the best and used them as the parents of the next generation. The process was continued, perhaps through hundreds of generations, until satisfactory results were achieved. Using this methodology, the program was able to create 14 circuits whose results are competitive with human designs. Ten infringed on existing patents with some surpassing the results available with these patents, and a few resulted in circuits that produced results previously thought impossible to achieve by experienced electrical engineers.

Given these results, it is tempting to speculate on future scenarios implicit in Koza's challenge to create a program that can produce human-competitive results. Imagine a computer dialing up the patent office and submitting its design electronically. When the patent is approved, the computer hires a lawyer to structure a deal with a company that produces components using the patented circuits, specifying that the royalties be deposited in its bank account, from which it electronically pays its electric bill. Or suppose that the computer submits an article describing its creation to an electrical engineering journal, using its serial and model numbers as the author. These science fiction scenarios aside, it is clear that Koza's results enable a serious case to be made for attributing to his genetic programs the human attributes of creativity and inventiveness. If one objects that the programs are 'dumb' in the sense that they do not know what they are doing and their designs are simply the result of blind evolutionary processes, one risks the riposte that humans also do not know what they are doing (otherwise they could describe explicitly their methods for solving these problems) and that their ability to solve these complex processes are also the result of blind evolutionary processes.[5]

Are Humans Special?

The idea that humans occupy a unique position in the scheme of things continues in the new millennium to be a pervasive and historically resonant belief. One of its contemporary defenders is Francis Fukuyama (2002), who argues for the proposition that 'human nature' exists and, at least in broad outline, can be specified as attributes statistically distributed along a bell curve. Among these attributes are the desire to care for one's children, the desire to favor one's kin group, the desire of males to have sex with females of reproductive age, and the propensity of young males for aggressive confrontations. Furthermore, he defends this human nature as the natural basis for social, cultural, and political institutions, arguing that those institutions that accommodate human nature will be more stable and resilient than those that do not. Finally, he argues that we must at all costs defend our human nature from technological interventions, outlawing or regulating practices that threaten to mutate and transform it in significant ways. He thus positions himself explicitly in opposition to researchers such as Hans Moravec and Ray Kurzweil. Although he does not mention Rodney Brooks (*Our Posthuman Future* was published the same year as Brooks' *Flesh and*

Machines), it is fair to say that he would resist some of Brooks' ideas, particularly the notion that a robot like Cog could be made to develop human-like attributes. The one characteristic that Fukuyama's argument leads him increasingly to privilege as he goes through the list of what computers can and cannot do is emotion. Machines, he acknowledges (2002: 168), 'will probably be able to come very close' to duplicating human intelligence, but 'it is impossible to see how they will come to acquire human emotions.'

In pursuing this argument, Fukuyama makes some strange moves. For example, he draws heavily on evolutionary theory to explain how 'human nature' was created, but he also cites with approval Pope John Paul II's assertion that although evolution can be seen as consistent with Catholic doctrine, one must also accept that by mysterious means, at some point in the evolutionary process souls are inserted into human beings. Fukuyama opines:

> The pope has pointed to a real weakness in the current state of evolutionary theory, which scientists would do well to ponder. Modern natural science has explained a great deal less about what it means to be human than many scientists think it has. (2002: 161–2)

But if one is free to suppose that human nature can be radically altered by supernatural forces other than evolution, why does evolution force us to conclude that human nature must be such and so? It seems that Fukuyama uses evolutionary reasoning when it is convenient for his argument and dispenses with it when it threatens his conclusion that human beings are special (in particular, that they have souls when no other living organisms do). This contradiction exposes the tautological nature of his argument. Humans are special because they have human nature; this human nature is in danger of being mutated by technological means; to preserve our specialness, we must not tamper with human nature. The neat closure this argument achieves can be disrupted by the observation that it must also be 'human nature' to use technology, since from the beginning of the species human beings have always used technology. Moreover, technology has coevolved throughout millennia with human beings and helped in myriad profound and subtle ways to make human nature what it is.

Another reason to view Fukuyama's argument with skepticism is the fact that the 'human beings are special' position has a long history that achieves special resonance in the triad of humans, machines, and animals. Although the configuration of these three terms has changed over time, the desire to arrange them to prove that humans are special has remained remarkably consistent. During the Renaissance humans were thought to be special because, unlike animals (who were then the closest competitors for the ecological niche that humans occupied), humans are capable of rational thought, a functionality they share with the angels and proof that they are made in God's image (at least men were; it took some centuries before women were admitted into the charmed circle of rational beings). As computational technology rapidly developed in the 20th century, it became more difficult to maintain that computers could not think rationally. The emphasis then shifted, as it does in Fukuyama, to the human capacity to feel emotions. Now it is animals with whom we share this functionality, it is denied to machines (not coincidentally, machines have in the meantime become the strongest competitors for the ecological niche that humans occupy). The ironies of this historical progression are brilliantly explored in Philip K. Dick's *Do Androids Dream of Electric Sheep?* ([1968] 1982). While the religion of Mercerism sanctifies the ability of humans and animals to feel empathy, the androids who so closely resemble humans that only the most sophisticated tests can tell them apart can be freely slaughtered and used as slaves because, allegedly lacking empathy, the essential characteristic of 'human nature', they are denied the legal protections given to humans and animals.

Although Brooks does not cite Fukuyama, he is of course familiar with the argument that humans are special. In paired sections Brooks (2002) considers the argument 'We are special' and then refutes it in 'We are not special'. His argument is so blatantly tautological that one would suspect him of making fun of Fukuyama (except it is unlikely that he had read Fukuyama's book at the time he was writing *Flesh and Machines*). The proof that machines can be made to feel emotions, he maintains in an argument that can scarcely be anything other than tongue-in-cheek, is that humans already are machines. Since humans feel emotion, it must therefore be possible for machines to feel emotion. Cutting through the Gordian knot of human nature by presupposing what is to be proved, the argument cannot be taken seriously. Nevertheless, it serves as an apt conclusion to the implications of the act–sense–think triad, as we shall see.

In Brooks' research acting becomes the attribute that humans and intelligent machines share. This convergence has the effect of emphasizing direct action in the world as a source of cognition rather than the neocortex. Consciousness, which no one has yet succeeded in creating in machines, is dethroned from its preeminent position and relegated to an epiphenomenon. Brooks' search for 'the juice' (2002) further positions human nature as convergent with machine intelligence. The force of this move can be seen in the conjunction of arguments proposed by Roger Penrose (1989, 1994) and Ray Kurzweil (1999), respectively. Penrose hypothesizes (without any evidence) that consciousness is a result of quantum computing in the brain, whereas Kurzweil is confident that quantum computing will allow computers to think as complexly as do human brains. Since all quantum computers can do at present is add one plus one to get two, everything in these arguments depends on future projections far ahead of what is currently the case. The future is thus crucial in allowing human beings to be understood as convergent with intelligent machines. The effect *in the present* is to emphasize those aspects humans share with intelligent machines, while those aspects of human being which machines do not share are de-emphasized. Fukuyama (2002) tries to reverse these priorities, emphasizing the parts of human nature that machines do not share, particularly emotions. In either case, human nature is understood in relation to intelligent machines.

Similar results emerge from the sensing and thinking apices of the STA triangle. Sensing becomes a research frontier not because we are likely to acquire new sensors through biological evolutionary processes; rather, sensing can be understood as part of our technological evolution and therefore an appropriate area for research. Programs that emphasize, as Cariani (1998) and Bird and Layzell (2002) do, the need for epistemic autonomy implicitly equate the evolution of artificial sensors with human agency and autonomy. Only when our machines can break free of the worldview embedded in the data we give them as input, they argue, can the power of autonomous perception really be freed from human preconceptions and thus become capable of yielding truly novel results. When machines are free to parse the world according to their autonomous perceptions, it remains an open question, of course, whether they will still follow agendas consistent with human needs and desires.

Cariani (1991: 789) acknowledges this possibility when he notes that epistemic autonomy is closely related to motivational autonomy:

> Such devices would not be useful for accomplishing our purposes as their evaluatory criteria might well diverge from our own over time, but this is the situation we face with other autonomous human beings, with desires other than our own, and the dilemma faced by all human parents at some point during the development of their children.

The rhetoric here moves in two directions, a configuration often seen in convergence scenarios. On the one hand, intelligent machines are envisioned as our children, a position that evokes sympathetic nurturing and empathic identification. On the other hand, they are also postulated as having their own goals distinct from those of humans, a prospect that Hans Moravec invokes when he talks about 'domestic' and 'wild-type' robots. While 'domestic' robots will be harnessed to tasks driven by human desires (like the intelligent vacuum cleaner that Rodney Brooks has developed and is currently marketing), the 'wild-type' robots are envisioned as being used for off-world exploration, mining, and other tasks, a mandate that makes it likely they will be given the capacity to reproduce, mutate, and evolve through automated off-world factories that have the potential to re-program themselves. Moravec postulates that the 'wild-type' robots will then evolve independently of humans, with the expectation that they will be free to define their own goals and desires – a situation startlingly like that anticipated in Philip K. Dick's book ([1968] 1982), with its satiric portrait of a dying human race that commits the hubristic sin of denying intelligent and self-aware androids the rights they so clearly deserve. Another way in which convergence scenarios aim to remove the anxiety associated with robots becoming our evolutionary successors is to imagine that the difference between humans and robots will become moot in an age when all intelligent beings can choose whatever embodiment they prefer, as Kurzweil (1999) does.

Working by other means, John Koza's genetic programs and their 'human-competitive results' point to similar conclusions. Now it is not merely rational thought that intelligent machines are seen to possess, but creativity and intuition as well. The fact that the programs arrive at these results blindly, without any appreciation for what they have accomplished, can be ambiguously understood as indicating that machines are capable of more creativity than that with which they have been credited, or that human intuition may be more mechanical than we thought. The idea that human intuition may originate not primarily in the conscious mind but through individual agents running their semi-autonomous programs, a proposition for which Marvin Minsky (1988) eloquently argued, makes Freud's version of the unconscious look hopelessly anthropomorphic by comparison. In this view, Freud's unconscious comes to seem like what the conscious mind imagines the unconscious would be rather than what it is, which is something at once more mundane and more alien – mindless programs running algorithms that are nevertheless capable of producing very sophisticated results, including a conscious mind that attributes these results to its own powers of rational thought.

Evaluating these diverse positions is akin to steering between Scylla and Charybdis. Certainly Brooks makes an important point when he critiques Kurzweil and Moravec for seriously underestimating the importance of human embodiment and its differences from the silicon instantiations of intelligent machines. Significantly, neither Kurzweil nor Moravec is trained in neurophysiology. Researchers with this experience, such as Antonio Damasio (1995, 2000), have given far different accounts of the complexities of human embodiment and the recursive feedback loops that connect brain to viscera, thought to emotion, consciousness to the specificities of humans. On the other hand, Brooks almost certainly underestimates the importance of consciousness in human culture and society. From the viewpoint of physical anthropology, consciousness may indeed be a late evolutionary add-on, but it is also the distinctive characteristic that has co-evolved with and is inseparable from the uniquely human achievements of language and the development of technologies.

While Brooks joins Kurzweil and Moravec in using the future to anchor their visions of human nature, Fukuyama wants to anchor human nature in the past, specifically the history of human evolution. In my view it is not possible and probably not desirable to constrain scientific research through the kind of legislative fiat Fukuyama advocates. 'Human nature' may indeed have an evolutionary basis, as he argues, but cultural and technological evolution have now so converged with biological evolution that they can no longer be meaningfully considered as isolated processes. Whatever our future, it will almost certainly include human interventions in biological processes, which means that 'human nature' will at least in part be what humans decide it should be.

Although these research programs have their own agendas and should not be conflated with one another, they have in common positing the intelligent machine as the appropriate standard by which humans should understand themselves. No longer the measure of all things, man (and woman) now forms a dyad with the intelligent machine such that human and machine are the measure of each other. We do not need to wait for the future to see the impact that the evolution of intelligent machines has on our understandings of human being. It is already here, already shaping our notions of the human through similarity and contrast, already becoming the looming feature in the evolutionary landscape against which our fitness is measured. The future echoes through our present so persistently that it is not merely a metaphor to say the future has arrived before it has begun. When we compute the human, the conclusion that the human being cannot be adequately understood without ranging it alongside the intelligent machine has already been built into the very language we use.

The crucial point suggested by my analysis is simply this: our future will be what we collectively make it. Future projections should be evaluated not from the perspective of how plausible they are, for that we cannot know with certainty, nor in the inertia of our evolutionary past, for that alone is not sufficient to determine what we can or will be. To accept the gambit of positioning the argument in either of these terms is already to concede the game to those who would hold the present hostage to the future or the past.

Rather, we should ask another kind of question altogether: what do we want the future to be? What values should command our allegiances and the commitment of our senses, thoughts, and actions? Viewed from this perspective, the STA paradigm serves to focus another sort of inquiry, leading not to the foregone conclusion that intelligent machines are the inevitable measure by which the human will be understood but rather to debates about how we can best achieve the future we want. Neither future projection nor past history is sufficient to answer this question definitively, and neither should be allowed to foreclose the urgency of grounding our future in ethical considerations. It is likely that our future will be increasingly entwined with intelligent machines, but this only deepens and extends the necessity for principled debate, for their futures too cannot be envisioned apart from the primary concern for ethics that should drive these discussions. What it means to be human finally is not so much about intelligent machines as it is about how to create just societies in a transnational global world that may include in its purview both carbon and silicon citizens.

Acknowledgements

I am indebted to Carol Wald for assistance with bibliographic research for this article, and especially for the link to William Gibson.

Notes

1. For extended critique of Kurzweil's predictions and his rejoinders, see Jay Richards (2002). Of special interest in the Richards collection, in view of my argument here, is Thomas Ray (2002: 116–27) and Michael Denton (2002: 78–97).

2. It is interesting that science fiction writers, traditionally the ones who prognosticate possible futures, are increasingly setting their fictions in the present. In a recent interview, William Gibson commented on this tendency. Andrew Leonard (2003) asks, 'Was it a challenge to keep writing about the future, as the Internet exploded and so much of what you imagined came closer?' Gibson replies:

> I think my last three books reflected that. It just seemed to be happening – it was like the windshield kept getting closer and closer. The event horizon was getting closer . . . I have this conviction that the present is actually inexpressibly peculiar now, and that's the only thing that's worth dealing with.

3. This article picks up where I left off in my book (Hayles, 1999), which explores the developments in changing constructions of the human in relation to intelligent machines from post-Second World War to 1998.

4. For a trenchant critique of these experiments, see Daniel Dennett (2003: Chapter 8). He argues the widespread interpretation of these experiments as indicating that 'choice' is an epiphenomenon is a result of what he calls the Cartesian Theater. The metaphor of Cartesian Theater mistakenly assumes that there is some central coordinating agency, some 'place' where consciousness resides and makes its decisions. Dennett (1992), by contrast, argues for a draft and revision model of consciousness that sees consciousness as a multilayered process in which many time-streams participate. Thus, there is no single 'time t' at which a decision is made but rather a multiplicity of times sampled at different points in the stream.

5. The implication that human creativity also operates like the computer programs is slyly intimated in *Genetic Programming III*'s dedication, which reads 'To our parents – all of whom were best-of-generation individuals,' the phrase used to select the winning circuit designs that will become the parents of the program's next generation.

References

Bear, Greg (2004) *Darwin's Radio*. New York: Del Ray.

Bird, Jon and Paul Layzell (2002) 'The Evolved Radio and its Implications for Modeling the Evolution of Novel Sensors', *Proceedings of Congress on Evolutionary Computation* (n.v.): 1836–41, http://www.hpl.hp.com/research/bicas/pub-10.htm (accessed 20 April 2003).

Bird, Jon and Andy Webster (2001) 'The Blurring of Art and Alife', http://www.cogs.susx.ac.uk/users/agj21/ccrg/papers/TheBlurringofArtandALife.pdf (accessed 20 April 2003).

Brooks, Rodney A. (1999) *Cambrian Intelligence: The Early History of the New AI*. Cambridge, MA: MIT Press.

Brooks, Rodney A. (2002) *Flesh and Machines: How Robots Will Change Us*. New York: Pantheon.

Cariani, Peter (1991) 'Emergence and Artificial Life', pp. 775–97 in C.G. Langton,

C. Taylor, J.D. Farmer and S. Rasmussen (eds) *Artificial Life II, SFI Studies in the Sciences of Complexity*, vol. X. Boston, MA: Addison-Wesley.

Cariani, Peter (1998) 'Epistemic Autonomy through Adaptive Sensing', pp. 718–23 in *Proceedings of the 1998 IEEE ISIC/CRA/ISAS Joint Conference*. Gaithersburg, MD: IEEE.

Damasio, Antonio (1995) *Descartes' Error: Emotion, Reason, and the Human Brain*. New York: Avon.

Damasio, Antonio (2000) *The Feeling of What Happens: Body and Emotion in the Making of Consciousness*. New York: Harvest.

Dennett, Daniel (1992) *Consciousness Explained*. New York: Back Bay Books.

Dennett, Daniel (2003) *Freedom Evolves*. New York: Viking.

Denton, Michael (2002) 'Organism and the Machine: The Flawed Analogy', pp. 78–97 in Jay Richards (ed.) *Are We Spiritual Machines? Ray Kurzweil vs the Critics of Strong AI*. Seattle, WA: Discovery Institute Press.

Dick, Philip K. ([1968] 1982) *Bladerunner* (original title *Do Androids Dream of Electric Sheep?*). New York: Ballantine Books.

Fukuyama, Francis (2002) *Our Posthuman Future: Consequences of the Biotechnology Revolution*. New York: Farrar, Straus and Giroux.

Haggard, Patrick and Benjamin Libet (2001) 'Conscious Intention and Brain Activity', *Journal of Consciousness Studies* 8: 47–63.

Hayles, N. Katherine (1999) *How We Became Posthuman: Virtual Bodies in Cybernetics, Literature, and Informatics*. Chicago, IL: University of Chicago Press.

Koza, John R., Forrest H. Bennett III, David Andre and Martin A. Keane (1999) *Genetic Programming III: Darwinian Invention and Problem Solving*. San Francisco, CA: Morgan Kaufmann Publishers.

Kurzweil, Ray (1999) *The Age of Spiritual Machines: When Computers Exceed Human Intelligence*. New York: Penguin.

Leonard, Andrew (2003) 'Nodal Point: Interview with William Gibson', *Salon.com* http:www.salon.com/tech/books/2003/02/13/Gibson/index.html (accessed 20 April 2003).

Libet, Benjamin (1985) 'Conscious Cerebral Initiative and the Role of Conscious Will in Voluntary Action', *Behavioral and Brain Sciences* 8: 529–66.

Menzel, Peter and Faith D'Aluisio (2000) *Robo Sapiens: Evolution of a New Species*. Cambridge, MA: MIT Press.

Minsky, Marvin (1988) *Society of Mind*. New York: Simon and Schuster.

Moravec, Hans P. (1990) *Mind Children: The Future of Robot and Human Intelligence*. Cambridge, MA: Harvard University Press.

Moravec, Hans P. (1999) *Robot: Mere Machine to Transcendent Mind*. New York: Oxford University Press.

Penrose, Roger (1989) *The Emperor's New Mind: Concerning Minds and the Laws of Physics*. New York: Oxford University Press.

Penrose, Roger (1994) *Shadows of Mind: A Search for the Missing Science of Consciousness*. New York: Oxford University Press.

Ray, Thomas (2002) 'Kurzweil's Turing Fallacy', pp. 116–27 in Jay Richards (ed.) *Are We Spiritual Machines? Ray Kurzweil vs the Critics of Strong AI*. Seattle, WA: Discovery Institute Press.

Richards, Jay (ed.) (2002) *Are We Spiritual Machines? Ray Kurzweil vs the Critics of Strong AI*. Seattle, WA: Discovery Institute Press.

Turkle, Sherry (1984) *The Second Self: Computers and the Human Spirit*. New York: Simon and Schuster.

N. Katherine Hayles is Professor of Literature in the English Department of the University of California, Los Angeles, teaches and writes on the relations of literature, science and technology in the 20th and 21st centuries. Her recent book *How We Became Posthuman: Virtual Bodies in Cybernetics, Literature, and Informatics* (University of Chicago Press, 1999) won the Rene Wellek Prize for the Best Book in Literary Theory for 1998–9. Her latest book, *Writing Machines*, won the Susanne Langer Award for Outstanding Scholarship. Her forthcoming book, *My Mother Was a Computer: Digital Subjects and Literary Texts*, will appear in September 2005 (University of Chicago Press).

Metamorphoses
The Myth of Evolutionary Possibility

Sarah Kember

Introduction

WITHIN A broad cultural context involving the convergence of the biological and the technological, one key feature has been the resurgence of Darwinism. Where the particular characteristics of this Darwinism have been vigorously contested (see, for example, Brown, 1999 and Rose and Rose, 2000),[1] a tacit belief in evolution *per se* (Midgley, 2002) has been at the root of key developments across the boundary of art and science. This article challenges the belief in evolution which has increasingly erased the boundaries between art and science in the particular contexts of artificial life and transgenic engineering. Here, more than elsewhere, there is an almost overwhelming sense of evolutionary possibility, and here more than elsewhere the evolution in question is more technological than biological, more informational than material, more possible than actual. Without seeking to reinforce or to map these dichotomies on to each other, my point is to show how, in these contexts, the complex interplay or dialogue between them is effaced, creating not just a tacit, but a sterile belief in evolution as an abstract process.

The critique of abstract evolution which follows is not a critique of evolution as a whole. Rather, it is part of a struggle which is both internal and external to computer science and to biological and evolutionary theory over the prioritization of form versus matter and over the epistemology and ontology of information.[2] My argument is not that evolution has no role in contemporary life and its various modes of interpretation but that it has been displaced, abstracted and somewhat reified through its attachment to the gene as the fundamental informational unit of life. In the world of abstract evolution, which seems to me to persist even in the face of contestation, it is principally genes which evolve and information which lives. My argument is, then, part a broader defence (Oyama, 1985; Keller, 1995; Haraway, 1997; Hayles, 1999) of the role of the biological *within* the technological, the material *within* the informational, the actual *within* the possible.

The Possible and the Actual

> Whether in a social group or in an individual, human life always involves a continuous dialogue between the possible and the actual. A subtle mixture of belief, knowledge, and imagination builds before us an ever changing picture of the possible. (Jacob, 1982: vii)

The terms of my critique of evolution are drawn from François Jacob's essays on modern biology and specifically on neo- or genetically informed Darwinism in *The Possible and the Actual*.[3] Though the terms themselves are not defined or interrogated,[4] they describe, respectively, human attempts to represent and to effect radical long-term biological transformations and the brakes or limitations imposed on them by science, politics, ethics and the imagination. The key point for Jacob is not so much to anchor the possible and the actual to demarcated realms such as myth and science, imagination and reality but to highlight the dynamic between them as it operates, or in the context of neo-Darwinism, fails to operate within and across such realms. The dynamic is that of containment through dialogue and there is or should be a dialogue between the actual and possible across a spectrum of belief, knowledge and imagination. Moreover, this dialogue, by containing the possible within the actual, balances radical and conservative visions of the evolution of biological forms, allowing space for those forms to 'really' change.

For Jacob, the picture of the possible – whether it be 'dogs with fish heads' drawn in 16th-century zoology books or aliens with two legs, two arms and two antennae drawn in 20th-century science fiction comics – has historically been limited by a dialogue with the actual. Monsters, aliens and hybrids have always been strangely familiar; a means of working out 'our desires and fears' (1982: viii) about future or other lives. In different ways and with different rules, politics, art and science are all ways of conducting this dialogue:

> For science, there are many possible worlds; but the interesting one is the world that exists and has already shown itself to be at work for a long time. Science attempts to confront the possible with the actual. It is the means devised to build a representation of the world that comes ever closer to what we call reality. (Jacob, 1982: 12)

The problem for Jacob is that science, through the influence of modern Darwinism with its concentration on the evolution and engineering of genes is focusing on the possible at the expense of the actual and so is losing (its) ground. It becomes clear in his writing that this lack of dialogue is of significance not just to the status of science but to the social realm and to questions of politics and ethics. This is the basis of my concern with the evolution of possible life *in vitro* and '*in silico*' (Langton, 1996: 51): there are, to state it rather bluntly, important political and ethical questions which are effectively sidelined in the hubris that harnessing evolution has generated. This problem, as I see it, is worsened by the equally unseemly rush to disavow responsibility, to look for the loophole in the Faustian law, by re-naturalizing the de-naturalized entities which emerge.

Where genetics, and especially transgenic engineering (the transfer of genetic and other biological material across species 'or even across taxonomic kingdoms, for example, from fish to tomatoes' [Haraway, 1997: 60]), is generating an industry and an imaginary of novel hybrid entities, the field of artificial life (alife) relocates the quest for artificial intelligence in the simulation and synthesis of computerized or robotic life forms (Kember, 2002). Through an examination of two works of art which address alife and transgenic engineering respectively – *Galápagos* by Karl Sims (1997) and *Genesis* by Eduardo Kac (1999a) – I want to suggest that the possible is not delimited by a dialogue with the actual in these contexts, but rather that it becomes part of what Jacob refers to as the over-extension, the mythologization of modern or neo-Darwinism. Possibility becomes totalized in the name of evolution; a means to no end or an end in itself. It becomes vacuous, empty of meaning, awaiting meaning. Possibility becomes its own pursuit, what Paul Virilio describes as a 'limit-performance' characteristic of a post-cold-war technoscience which has lost its purpose, lost its way (2000: 1). Technoscience, in this context fails to see beyond the endless scope of evolutionary possibility and the 'radiant future of transgenic species' (2000: 145), like the future of alife's artificial 'autonomous' agents (Maes, 1996, 1997) is an economic vision, a vision of the instrumentalization of life which this particular 'sci-art' fails to challenge.

Jacob's use of the concept of myth is salutary here. On the one hand, it refers to a sterile belief. Darwinism, like 'any theory of some importance is liable to be over-used and to slip into myth.' In explaining too much it

> ultimately explains very little. Its indiscriminate use invalidates its usefulness and it becomes empty discourse. Enthusiasts and popularisers, in particular, do not always recognise the subtle boundary that separates a heuristic theory from a sterile belief; a belief which, instead of defining the actual world, can describe all possible worlds. (1982: 22)

This is precisely the way in which (Darwinian) evolutionary possibility becomes mythical. Moreover, Jacob also refers to myth as 'a story giving the origins and therefore explaining the meaning and purpose of the living world as well as man's place in it' (1982: 23). The theory of evolution is one such story, but he also tells of how genetic engineering, by transgressing the laws of nature (and sacred culture) 'conjures up some of the old myths that have their origin in the deepest kind of human anxiety' and which deal with 'the primitive terror associated with the hidden meaning of hybrid monsters, the revulsion caused by the idea of two beings unnaturally tied together' (1982: 46). Metamorphoses is the 'old myth' which deals at once with evolutionary possibility and the human anxiety it can provoke.[5]

Metamorphoses, from the Greek *meta* (across, after or between) and *morphē* (form) is, as presented in a recent exhibition curated by Marina Warner and Sarah Bakewell[6] as much a facet of human nature as human culture.[7] It describes fundamental biological and evolutionary processes of both the self and the species, including the movement of bodies through birth, reproduction, ageing and death and the mutations and selections operating on them over considerable lengths of time (Warner and Bakewell, 2002: 2). As they see it, the complex and highly evolved human brain has produced and depended on stories, rituals, myths and art, a great deal of which 'return again and again to tales of transformation and creation' (Warner and Bakewell, 2002: 2). The curators of *Metamorphing: Transformation in Science, Art and Mythology* set out to universalize metamorphic nature/culture and maintain that 'mythologies all over the world feature the magical shape-shifting of gods and goddesses, and most great religions place metamorphic myths at the centre of their tradition.' So, for example, in Christianity, transformation can be divine or demonic, bread and wine into the body and blood of Christ or Satan's many disguises. In Hindu mythology, the first man and woman transform themselves in order to create all the species of the world (2002: 2). The tales of transformation are transformed across history as well as cultures 'gaining in depth and range' through the encounter with science 'and especially with the theory of evolution'. Here metamorphoses is naturalized; not a reference to shape-shifting but to the 'organic unfolding of forms in time and space' (2002: 2). Darwin's theory of evolution constitutes a set of 'simple metamorphic rules: descent with modification, selection pressure, and adaptation to environment' (2002: 12) and with the arrival of genetics these rules become 'even more defined, describing the replication and mutation of simple units of heredity' (2002: 12). Darwin's metamorphic rules are apparently so simple that they can be replicated in computer techniques such as morphing, in which one form transforms seamlessly into another through various intermediate stages. The exhibition included a computer artwork *Origin* (1999) by Daniel Lee. Via a sequence of images of mythical amphibians and mammals, Lee morphs or metamorphoses seamlessly from a shapeless originary form to the human form, reinforcing the play of the possible and actual and the conflict between open-ended evolution – 'chance' – and the tree of life (with the human species securely at the summit) – 'necessity' (Monod, 1974). This conflict concerning the status of man in relation to animal reveals itself in evolutionary theory through the contested currency of what Mary Midgley calls the 'escalator model', the Lamarckian idea that 'evolution is a steady, linear upward movement, a single inexorable process of improvement' leading to man 'and beyond into some superhuman spiritual atmosphere' (Midgley, 2002: 7). Though she is keen to contrast this with Darwinian theory, Midgley maintains that there is more than a residual sense of divine order in secular and scientific accounts of metamorphoses from the theory of evolution to various manifestations of modern Darwinism such as artificial life and genomics. This may be said to account for what Warner describes as the 'terrors and pleasures' which attend 'any technological change which undermines our humanity and identity by merging our bodies with animals or with machines, or by taking away our sense of having a unique, unchangeable soul – even when we understand that sense to be illusory' (Warner, 2002: 3).

Ovid's epic poem consists of many stories concerned with the transmigration of souls and the transformation of bodies – 'All things are always changing, but nothing dies' (in Warner, 2002: 1) – but Pythagoras' vision of interconnectedness and continual flux (stemming from classical and eastern thought and subsequently developed through biology and evolutionary theory) is, as Warner argues, inconsistent with a sense of perfection and of the stabilization or fixity of forms in the poem: 'when Daphne is turned into a laurel tree and a young man called Cygnus becomes a swan . . . the shape into which they shift more fully expresses them and perfects them than their first form' (Warner, 2002: 4). On the other hand, the punitive aspect of metamorphoses which recurs in Ovid (2002: 9) is underlined in the Christian imagery of Dante's *The Divine Comedy*. In the Christian imaginary each individual has a unique body and soul joined together by God (Warner and Bakewell, 2002: 4) so shape-shifting 'becomes a game played by demons' and a symbol of 'sin, terror, torment and loss of selfhood' (Warner and Bakewell, 2002: 4). Dante's Inferno was among the first re-tellings of Ovid and here, in hell, all metamorphoses are a form of torture and centre on the loss of self and/or species identity. Thus,

> in the circle of thieves Dante describes how a snake fastens on a sinner, intertwines his multiple limbs about him, and mingles so deeply with his being that his prey ceases to be an entity at all but merges into an 'imagine perversa' (a perverse image). (Warner and Bakewell, 2002: 36)

For Dante 'in the afterlife of the damned, morphing utterly reduces identity and integrity' and 'the punished lose their natures as their matter is changed, exchanged, transmuted' (Warner and Bakewell, 2002: 37). Ovid is thus subjected to Christian morality, a conservative revision, and through artworks such as Botticelli's 'Drawings to Dante' or Bosch's 'Garden of Earthly Delights' the affinity between transformation, damnation and the loss of identity is established (2002: 38) even, for Braidotti, as far as contemporary popular cultural representations of 'monstrous, mutant or hybrid others' (Braidotti, 2002: 5). So the metaphor of metamorphoses returns the contemporary vision of possible life to the realms of heaven and hell, perfection and punishment, drawing a familiar boundary around the process of transformation and the sanctity of human life.

In artificial life and transgenic engineering, what Jacob refers to as the 'sterile belief' in abstract evolutionary possibility recalls the 'old myth' of metamorphoses which, constrained by both God and Darwin, creationism and evolutionism, has nowhere new to go. Evolutionary possibility is effectively negated by the God-given sense that transformation is either transcendent or transgressive. Metamorphoses is a conservative mythology in which nothing actually changes, despite the centrality of change itself. This reading of metamorphoses as a conservative mythology does not seek to preempt Rosi Braidotti's appropriation of it as a more radical one, but rather to highlight the impotence of the evolutionism which it contains and supports and which shadows a quest for metamorphoses as 'becoming versus Being in its classical modes' (2002: 2).[8]

The conservative metamorphic myth of evolutionary possibility can be traced through the discourses of artificial life and transgenesis and is drawn out here through an analysis of two art works which are regarded as being indicative, though by no means exhaustive, of these fields. The point of exploring the boundary between art and science is to demonstrate the pervasive character of the myth in question and thereby a certain imaginative poverty in contemporary technoscientific culture.

Galápagos

Karl Sims's *Galápagos* (1997) is described as an interactive media installation which allows users to 'evolve' 3D animated forms.[9] Sims is an established alife artist whose work, including the previous *Co-Evolved Virtual Creatures* (1992) is part of a lineage which is traced back through Thomas Ray's *Tierra* (Ray, 1996) and Richard Dawkins's *Biomorph* (Dawkins, 1991) via William Latham's *Evolutionary Art* (Todd and Latham, 1992) to John Conway's *The Game of Life* developed during the 1960s.[10] Like most of this work, *Galápagos* seeks not to interrogate but to instantiate Darwinian evolution in the computer:

> Galápagos is an interactive Darwinian evolution of virtual 'organisms'. Twelve computers simulate the growth and behaviours of a population of abstract animated forms and display them on twelve screens arranged in an arc. The viewers participate in this exhibit by selecting which organisms they find most aesthetically interesting and standing on step sensors in front of those displays. The selected organisms survive, mate, mutate and reproduce. Those not selected are removed, and their computers are inhabited by new offspring from the survivors. The offspring are copies and combinations of their parents, but their genes are altered by random mutations. Sometimes a mutation is favourable, the new organism is more interesting than its ancestors, and is then selected by the viewers. As this evolutionary cycle of reproduction and selection continues, more and more interesting organisms can emerge. (Sims, 2003: 1)

The human/machine collaboration here seems to constitute an optimum metamorphic system (with the capacity for emergence or unpredictability)[11] and a means of combining meaning (as aesthetics rather than ethics)[12] and maths. Where 'the visitors provide the aesthetic information by selecting which animated forms are the most interesting', the computers 'provide the ability to simulate genetics, growth, and behaviour of the virtual organisms' with overall surprising results which neither could produce alone (Sims, 2003: 1). The 'double-barreled software-wetware approach is crucial' (Frauenfelder, 2003: 1) to this digital display of Darwinian evolution which is ultimately both the subject and the object of the installation, dependent on but displacing both humans and machines from the centre stage. What Sims seeks to capture is not a new and improved form of interactivity[13] but a kind of Darwinian hyperspace, the hyperspace of all possible organisms which could/can exist despite the limitations of either human or machine. What is important is that 'the results can potentially surpass what either human or machine could produce alone' (Sims, 2003: 1). This is an ideal space of evolutionary possibility, uncontaminated by external factors – a direct digital analogue of the isolated Galápagos Islands where Darwin was able to observe 'a rare example of a relatively independent evolutionary process' (Sims, 2003: 2). Sims's ultimate aim is to present, through simulation, evolution 'comparable to the value that Darwin found in the mystical creatures of the Galápagos Islands' (2003: 2).

Sims's mystical creatures 'possess both the variety and the structural similarity of biological organisms' (Unger, 2003: 1) and despite the suggestion of human (aesthetic-eugenic) intervention (Unger, 2003) their transformation and evolution are seemingly open-ended and guaranteed by the process of genetic combination and mutation. Miles Unger describes how

> in my quest for intriguing new progeny I can combine the characteristics of different creatures by stepping on a few footpads in rapid sequence. After a few passes, the fauna is radically altered as robotic contraptions give way to slithering jellyfish fronds. (2003: 1)

The subject of this process, the 'I' is in effect indistinguishable from the object. Unger, in combination with the computers, performs the role of simulated evolution which is itself agential over and above not only the software-wetware infrastructure but the rapidly evolving increasingly complex life-forms themselves. These only achieve fixity, perfection or finality retrospectively and through the provision of a genealogy file ('Previous') which contains creatures from previous generations (Sims, 2003). Alongside the genealogy option is the seemingly ever present option to 'Start Over', to return to the origin of life and cause 'a new evolution to begin from scratch with simple randomly generated creatures'. Also omnipresent in alife programming is the belief, the faith that 'each new evolution will generate results that have never been seen before' (Sims, 2003: 1). Again though, it is not the results themselves, not the mystical kinds and creatures analogous to Darwin's finches but the capacity to produce them which is paramount. The possible exists here at the expense of the actual. As Unger states, the success of *Galápagos* 'lies not in its forms, which are ever-changing and unpredictable, but in its architecture – the algorithms, or set of computational rules, that generate those forms'

(2003: 2). The real beauty, the real meaning, is the maths and its capacity to mimic a simplified, deterministic relationship between genotype and phenotype – over and over again at very high speed: 'In *Galápagos*, the process itself is the real art. Sims has created a remarkable allegory of the mechanism that four and a half billion years ago began the most complex design of all – life' (Fifield, 2000: 2).

In *Galápagos*, evolution = art = life itself, but this is a simulated evolution, a virtual evolution, thoroughly informationalized and abstracted from the corporeal and material realms of fleshy bodies in local environments. *Galápagos* installs in an artwork Langton's foundational theme that 'life is a property of form not matter', a 'kind of behaviour, not a kind of stuff' (1996: 53). For him, 'just as the "logical form" of a machine can be abstracted from its material substrate, so too . . . may the "logical form" of a [living] organism . . . be separated from its material of construction' (in Jonson, 1999: 47). So logical form – 'information epitomised in the digital code or program' (Jonson, 1999: 48) – is itself 'ontologised as life's disin-carnate essence and origin. Concomitantly, material substance is styled as code's inessential, merely accidental or secondary supplement' (Jonson, 1999: 49). Matter, in other words, is superfluous to life and to the evolution of life. The elision of matter, contested both within the subsequent field of alife and beyond (in philosophy and feminism, for example) nevertheless has, as Jonson points out, 'a distinguished scientific provenance' within molecular biology and its focus on the gene as the fundamental informa-tional agent of life. The information contained in the genetic code or program, as outline in Francis Crick's Central Dogma, determines the development of proteins in the cells of the body, and this information crucially flows only one way. There is, then, as critics of the Central Dogma point out, no acknowledgement of environmental influence – whether intra-or extra-cellular – in the development of the organism; no sense of the relation between information and matter in the constitution of life. This vision of the organism and of (a)life is perpetuated in *Galápagos*, hooked as it is on the eternal, unbounded possibility of logical form, gene agency, evolution and aesthetics without reference to the actual circumstances of the organism, matter, human/machine environments and ethics. The environment of simulated evolution is global, hermetic and homogenized. What matters here is the self-organizing, self-replicating system itself and not its component parts or creatures or kinds. There is no real difference (between one kind and another, between art and instrumentalism), no real distance (in time or space) and only one truly viable organism; the 'simpli-fied evolutionary system' which can be 'observed from start to finish and run multiple times' (Sims, 2003: 2).

> The matter-time of the hard geophysical reality of places gives way to this light-time of a virtual reality which modifies the very truth of all durée, thereby provoking, with the time accident, the acceleration of all reality: of things, living beings, socio-cultural phenomena. (Virilio, 2000: 117)

The acceleration of all reality yields authority to the machines and those who programme them (Virilio, 2000: 122), those for whom the primary pursuit would appear to be more aesthetics than ethics, more progress than politics; a 'limit-performance' (Virilio, 2000: 1) at the boundary of art, science and technology. Moreover, the acceleration of all reality leads, for Virilio, inexorably towards inertia:

> Every time we introduce an acceleration, not only do we reduce the expanse of the world, but we also sterilize movement and the grandeur of movement by rendering useless the act of the locomotor body. Similarly, we lose the mediating value of 'action' while that of the immediacy of 'interaction' gains in comparison. (Virilio, 2000: 123)

In *Galápagos*, ultimately, nothing evolves but evolution itself; the system speeds along, accelerating evolution in the abstract, producing molecular metamorphoses, strangely familiar forms held within the ultra-Darwinian hyperspace of all possible organisms-as-genes. The animated display of mutating virtual entities in *Galápagos* re-captures the theme of metamorphoses as a myth of possibility: 'all things are always changing, but nothing dies.' Rather, 'those not selected are removed, and their computers are inhabited by new offspring from the survivors' (Sims, 2002: 1). Artificial selection by aesthetic criteria substitutes for natural selection by criteria of fitness understood as being contingent on, relative to a given environment. It is by eliding all environmental influence – not just organic matter – that *Galápagos* invests in the possible at the expense of the actual. There is no context for or containment of possibility when even the role of the user and viewer let alone that of human/machine interaction is effaced by the overriding sovereign subject of evolution in the abstract. By being uncontained, abstract evolutionary possibility is simply over-extended – impotent, sterile, mythical in Jacob's senses of the word. It goes nowhere fast and produces nothing new. Every 'organism' is recognizable at least in part – robot-like, crab-like, or with jellyfish fronds – and its evolution is compromised in that it is partly designed or selected by the human eye, by human knowledge and belief. What Ovid's *Metamorphoses* still tells us is that, faced with apparently unlimited possibility, humans have a tendency to be quite conservative. There is more than a hint of the quest for perfection in the genealogy file and even of the recourse to punishment in the erasure of the relatively un-aesthetic. The organisms themselves, after all, do not matter and for me it is this which underlies the lack of dialogue and imagination in this work.

Genesis

Eduardo Kac's genetic, or as he terms it 'transgenic art' is comparable to Sims's work in as far as it locates the origin and evolution of life (the beginning and the end) within a reified molecule,[14] and in *Genesis* (1999)[15] Kac also enacts the replacement of God with evolution in the abstract.[16] Genesis is described as 'a transgenic artwork that explores the intricate relationship between biology, belief systems, information technology, dialogical inter-action, ethics, and the Internet' (Kac, 2002: 1). Such a wide exploration is focused through/on 'the key element' of the work; an 'artists' gene'. This is a synthetic gene which Kac generated by 'translating a sentence from the biblical book of Genesis into Morse Code[17] and converting Morse Code into DNA base pairs' according to a 'conversion principle' (Kac, 2002: 1). The sentence in question – 'let man have dominion over the fish of the sea, and over the fowl of the air, and over every living thing that moves upon the earth' – was chosen 'for what it implies about the dubious notion of divinely sanctioned humanity's supremacy over nature'. Kac seeks to challenge the originary hierarchy of animal/human, nature/culture and the Word of God through a collective interactive process of evolution-as-reinscription:

> The Genesis gene was incorporated into bacteria, which were shown in the gallery. Participants on the web could turn on an ultraviolet light in the gallery, causing real, biological mutations in the bacteria. This changed the biblical sentence in the bacteria. The ability to change the sentence is a symbolic gesture; it means that we do not accept its meaning in the form we inherited it, and that new meanings emerge as we seek to change it. (Kac, 2002: 1)

Kac places his faith in the transparency of genetic and/as natural language and in the information paradigm in molecular biology. His artwork refers to the Rosetta Stone and its three languages (Greek, demotic script, hiero-glyphics) as the 'key to understanding the past', constructing an analogy in the 'triple system of *Genesis* (natural language, DNA code, binary logic)' as 'the key to understanding the future' (Kac, 2003: 3). *Genesis* is clearly premised on the idea that 'biological processes are now writerly and programmable', and its end point is marked by decoding the altered biblical sentence, reading it back in plain English and establishing the 'insights' gained 'into the process of transgenic interbacterial communication' (Kac, 2003: 3). In fact, at the 'end of the show' of evolutionary reinscription, the original biblical sentence is only slightly undone, replaced not with the emergence of new meaning but with meaninglessness, nonsense.

Richard Doyle (1997) has written engagingly about what might be characterized as the endgame of molecular biology (Kember, 2002). The discovery of DNA as the secret of life, and particularly the map of the human genome may be said to mark the end of the story of life. Biology is faced with the realization 'that there is nothing more to say' (Doyle, 1997: 20). There is nothing at the heart of the organism which thus becomes a virtual object upon which a new story of information (as life) is based:

> In molecular biology, the end of the grand narrative of life, the 'death' of life, is overcome through a new story of information, in which a sequence of 'bits' is strung together or animated into a coherent whole through the discourse of 'that is all there is', a story of coding without mediation or bodies. (Doyle, 1997: 22)

With the advent of the human genome projects, secrets give way to sequences – of nucleic acids which are really 'all there is' (1997: 22). Life is displaced from the body and dispersed 'through the narratives and networks that make up the interpretations of genetic databases' (1997: 24). Information is a new sublime, matching the unrepresentable vision of life in the 19th century with 'the story of resolution told in higher and higher resolution' – that is 'the continual story that there is nothing more to say' (1997: 20). Genetic language succumbs to (a Beckettian) form of nihilism which is only enhanced by revelations of the unexpectedly low number of genes in the human genome and of the consequent increase in complexity or unintelligibility. What appeared, in 1953, to be a clear and legible code, now appears to be, in Steve Jones's terms, somewhat 'baroque'. Jones points out that the current realization that 'the working genes of higher organisms make up only a small proportion of their DNA' came as something of a shock to the founders of molecular biology, and he seems to substantiate the idea that genetic language is not only ambiguous (Kay, 2000) but also largely empty, redundant, even partially extinct and far from revelatory:

> Often genes themselves are interrupted by strings of bases that code for nothing. The whole sequence, discontinuous though it may be, is read off into RNA and – with a perversity alien to physics – edited to cut out redundant sections. Even worse, much of the DNA consists of repeats of the same sequence. A series of letters is followed by its mirror image, and then back to the original, thousands of times. Scattered among all this are the corpses of genes that expired long ago, and can be recognised as such only by their similarity to others that still function. The image of genetic material has changed. No longer is DNA a simple set of instructions. Instead, it is a desert of rigidity and waste mitigated by decay. (Jones, 1999: 4)

More Beckett than baroque, the death of genetic language informs the beginning of the end (the end of the beginning) of the story of life as told by and against molecular biology at the start of the 21st century.

Eduardo Kac's subsequent attempts to add a critique to his original gene-centred project[18] have much the same recuperative effect as attempts to add complexity to genetic determinism (Kember, 2002). Ultimately the doctrine of DNA prevails and his transgenic art instantiates another molecular metamorphosis where what matters is not the emergent forms or transgenic species, but the species independence of evolution in the abstract.[19] Kac employs two kinds of bacteria in *Genesis*: a kind which has incorporated a plasmid (extrachromosomal ring of DNA) containing ECFP (Enhanced Cyan Fluorescent Protein) and a kind which has incorporated a plasmid containing EYFP (Enhanced Yellow Fluorescent Protein). ECFP and EYFP are GFP (Green Fluorescent Protein) 'mutants'. Where ECFP bacteria contain the synthetic artists' gene, the EYFP do not. They emit cyan and yellow light respectively when exposed to UV radiation and where mutations occur naturally in the plasmids 'as they make contact with each other plasmid conjugal transfer takes place and we start to see colour combinations, possibly giving rise to green bacteria' (Kac, 2003: 2).

In the gallery, the central display – flanked by the sentence extracted from the book of Genesis on the right wall and the *Genesis* gene on the left – is a video projection of colour combination and the emergence of green bacteria. Via the projector, the petri dish of bacteria 'shows larger than life' on the centre wall – (extra)planetary in dimension, just like the earth seen from outer space, a symbol for Franklin et al. of globalization (2000). Both local and remote (web) participants are invited to engage in the process of (global) evolution, by turning the UV light on and off ('In the beginning . . .'). It is important to note, however, that this engagement is limited by Kac to 'monitoring', 'interference' and 'acceleration' of the evolutionary process: 'The energy impact of the UV light on the bacteria is such that it disrupts the DNA sequence in the plasmid, accelerating the mutation rate' (2003: 1).

Just prior to the creation of his now infamous 'GFP Bunny' (Alba, 2000) and subsequent to his plans for a green fluorescent dog (GFP K9), Kac has created a global vision of microbial transgenic life which usurps the meaning of life-as-we-have-known-it (Langton, 1996) and which is premised on mixing or miscegenation of categories (natural and artificial) and kinds in order 'to create unique living beings' (Kac, 1999: 289)[20] or new metamorphosed combinations. Within both the emergent art and industry of transgenesis, the utility and ubiquity of GFP in test case transgenic experiments have produced a global vision of life-as-it-could-be (Langton, 1996), and that vision is green (the colour, conventionally, of aliens). GFP is isolated from the Pacific Northwest jellyfish (*Aequoren Victoria*) and is, as Kac describes it 'species independent', needing no additional proteins or substrates for green light emission. Consequently, 'GFP has been successfully expressed in several host organisms and cells such as E. Coli, yeast, and mammalian, insect, fish and plant cells' (Kac, 1999: 290). So, at the outset of the global transgenic experiment, human kind is faced with the perhaps unsettlingly egalitarian prospect that 'every living thing that moves upon the earth' – including 'man' – has the potential to be green where green is not a pure colour, type, race or species. The nervousness which attends such a prospect is displayed in the inevitable jokes, puns and quips which attend each new green arrival – whether actual or possible. Where GFP K9 ('or "G", as I affectionately call it') will 'literally have a colourful personality' (Kac, 1999: 290), ANDi the transgenic monkey produced by the Oregan Regional Primate Research Centre (in October 2000) and in whom the GFP failed to take, was 'not quite a glowing success' (in Meek, 2001: 1). Jokes, as Haraway points out in her analysis of a primate research lab (1989), are a means of disavowing the flip-side of humanism in science: the sadism, racism, or sexism. They tell 'us' that mixing and miscegenation, like the failure rates and suffering induced in transgenic experiments, *is* in fact, species-specific and reassure us that 'we' will never be (made to be) 'green'. Human(ist) self and species identity is thus recuperated in both the art and industry of transgenesis (a truly – conservatively

– metamorphic praxis) and Kac's sentimental gestures towards kinship with his green dog/bunny – centring as they do on the 'domestication and social *integration* [my emphasis] of transgenic animals' (1999: 292) – are both hollow and an inadequate response to the increasing instrumentalization of life in the burgeoning transgenic industry of 'pharming'. Kac accepts, even embraces, the inevitability of transgenesis – 'it is clear that transgenesis will be an integral part of our existence in the future' – seeming to cancel out the danger of the farm with the promise of the expanded family and the progress of the (human) species:

> In the future we will have foreign genetic material in us as today we have mechanical and electronic implants. In other words, we will be transgenic. As the concept of species based on breeding barriers is undone through genetic engineering, the very notion of what it means to be human is at stake. However, this does not constitute an ontological crisis. To be human will mean that the human genome is not a limitation, but our starting point. (Kac, 1999b: 293)

Conclusion

Genesis, like *Galápagos*, participates in the metamorphic myth of possibility where nothing metamorphoses but metamorphoses itself. Metamorphoses is consistent with the theory of evolution where what is at stake is the mutation and transformation of both individual organism and species. Contemporary biotechnology, and particularly artificial life and transgenesis, at once denaturalize (accelerate) and re-naturalize this process of transformation, insisting on its status as evolution not becoming but eliding all of the material constraints within which evolution has operated and continues to operate (cells or computers, bodies or machines, communities or networks and so on). This is at the heart of Jacob's critique of modern Darwinism and his own attempts to retrieve the dialogue between the possible and the actual. There is corporeality in *Genesis* but no actuality; the mutation and evolution of the artist's gene are 'species independent', autonomous and unbounded by the organisms and their environment including the human participants who ultimately only 'monitor', 'interfere' or at most 'accelerate' it. The absence of the actual is ultimately what accounts for the conservatism, the humanism of Kac's vision. By over-extending evolutionary possibility he retains the spectre of the transgressive 'imagine perversa', the monster, the alien (the little green man?) which effectively precludes the transformation in human self and species identity which biotechnology promises or threatens. Unconstrained evolution in the abstract, the possible without the actual, as critics such as Jacob have made clear, changes nothing, let alone the order of things. Rather, it ensures the survival of God and Darwin in an unimaginative information loop (gene – Word, Word – gene) which has surely had its day.

Notes

1. The contest within modern Darwinism is between those who lay greater stress on the role of the organism and on the role of the gene within evolution. As Jacob (1982: 19) says, 'Virtually all biologists today believe in modern Darwinism. Some, however, think of evolution in terms of organisms, others in terms of molecules . . . there are still several possible ways of looking at evolution, its tempo and mechanism'. It is through the centrality of the gene, conceived of (within molecular biology) as the fundamental informational unit of life, that evolution becomes both abstracted and over-extended. Hence, it is the genetic centrality within evolutionary theory which this article is primarily concerned with.

2. The convergence between computer science and biology is exemplified in the discipline and discourse of artificial life. The founder of artificial life, Christopher Langton, argues that '[life] is a property of form not matter', reinstating, as Annemarie Jonson (1999) maintains, a fundamental Platonic distinction which is then highly contested both within artificial life and more widely. As a property of form not matter, the life in alife need not necessarily be organic (its criteria include: self-organization, self-replication, evolution and autonomy) and is primarily informational: 'Langton sees life as an abstract phenomenon, a set of vital functions implementable in various material bases' (Boden, 1996: 8). It consists of dynamic processes including 'many aspects of information processing, which he . . . sees as fundamental to life'. Boden is clear that 'this is no mere flight of hermeneutic fancy. For Langton, *information, communication*, and *interpretation* are real computational properties of certain formally describable systems'. What is more, 'these assumptions are widespread in the field'. It is important to add, as Boden does, that:

> one does not have to be a devotee of Langton, or even of A-Life, to share this view. Informational . . . concepts were widely used by theoretical biologists long before Langton gave A-Life its name. Familiar examples include the genetic 'code', with its attendant 'reading', 'interpreting', and 'transcription'; 'programs' of 'instructions' conveyed by specific cells or chemicals; 'messages' passed within and between cells; and cybernetic notions of 'informational feedback' and oscillating 'control'. (1996: 8)

The epistemological distinctions between matter and information and the ontology of information to which I refer are drawn predominantly from alife discourse, where, I am suggesting, they are foundational if by no means unproblematic or unproblematized (Jonson, 1999). Indeed, this problematic defines what Hayles refers to as the 'condition of virtuality' (Hayles, 2001).

3. What is at stake within modern Darwinism is Darwin's theory of evolution by natural selection as it has been interpreted since 1953 and the elucidation of the structure of DNA by Crick and Watson. This drew attention to the role of genetics in evolution where 'originally the theory of evolution was based on morphological, embryological, and paleontological data' (Jacob, 1982: 18). Modern or genetically informed evolutionary theory is referred to as neo-Darwinism but there is also an important scientific and ideological distinction drawn between neo-Darwinism and a 'fundamentalist' (Gould, 2000; Rose and Rose, 2000) form of ultra-Darwinism which focuses on the gene almost to the exclusion of the organism, which lays arguably more stress on the role of adaptation and natural selection than Darwin

himself did and which is strongly associated with Richard Dawkins's doctrine of the selfish gene (1976) and the development of sociobiology.

4. They are certainly not, as Jacob uses them, part of scholastic philosophy, but rather of his critique of modern Darwinism and the 'sterile belief' in evolution. There is thus no direct connection between philosophical concepts of the virtual and the actual (Deleuze, 1994; Lévy, 1998) and Jacob's concepts of the possible and the actual. Indeed, they may be said to describe two parallel trajectories within contemporary culture – becoming and evolution – which may be too easily obscured, and which this article, with its focus on evolutionary possibility seeks to clarify. The discourse of evolution, unlike that of becoming, is inherently natural-ized and Jacob's call for a dialogue between the possible and the actual is used here to effect a de-renaturalization of evolutionary possibility in contemporary art and science.

Even though Jacob's use of the terms does not map onto Deleuzian philo-sophical uses, it is interesting to consider Deleuze's concept of the possible as that which is not real (where the virtual is real in the sense that the past and memory are real) but which *may be* actual (where the virtual may not be): 'In other words, there are several contemporary (actual) possibilities of which some may be realised in the future' (Hardt, in Shields, 2003: 25). Jacob's 'dialogue' between the possible and the actual allows for this change (in the real), and it is this change in the real which is lacking in the sci-art discussed here (but which need not be). Simply put, in Jacob's critique of evolution in modern Darwinism, the possible is too far removed from the actual to lead to change. Rather, it leads nowhere except to abstract evol-utionary possibility itself: 'the *possible* is that which does not really exist, but could to various extents. At one extreme is the absolutely *abstract*, and an ideal which, properly speaking, has no existence, but rather only possibility' (Shields, 2003: 25).

5. There is also the play of myth and science in Jacob's essay of that title. Where 'in some respects at least' myths and science share a function of providing represen-tations of the world and of the forces that govern it; of fixing 'the limits of what is considered possible' (1982: 9); of turning chaos into order (1982: 11); of allowing space for the imagination and for the invention of 'a possible world, or a small piece of a possible world', the similarities end there.

> Having constructed what it considers not only as the best world but as the only possible one, it [myth] easily fits reality into its scheme . . . For scien-tific thought, in contrast, imagination is only one aspect of the game. At every step, it has to meet with criticism and experimentation in order to determine what might reflect reality and what is mere wishful thinking. (Jacob, 1982: 12)

6. *Metamorphing: Transformation in Science, Art and Mythology*, curated by Marina Warner and Sarah Bakewell was at the London Science Museum, 4 October 2002–16 February 2003.

7. Warner and Bakewell here maintain a distinction between nature and culture which is now widely questioned and which is certainly problematized in the areas addressed in this article.

8. It is not my intention to take issue with Braidotti's argument or to engage in detail with it here. Notions of evolution, with which I am concerned, do not map onto those of becoming, with which Braidotti is concerned, although they certainly

exist in a productive tension. My point is precisely to clarify, highlight and critique the existence of (abstract) evolutionism which informs and is informed by the current myth of metamorphoses and which in my view constrains but does not necessarily negate its association with a materialist theory of becoming.

9. *Galápagos* was installed at the ICC in Tokyo from 1997 to 2000, and was exhibited at the DeCordova Museum in Lincoln, Massachusetts, in 1999.

10. The whole field of artificial life was inspired by the work of John von Neumann (in the 1950s) and John Conway (in the 1960s) on cellular automata (CAs). This aimed to simulate biological processes of self-replication, evolution and emergence on a grid system or check board on which squares functioned as cells (Levy, 1992). The emergence of both stable and periodic configurations on Conway's *The Game of Life* grid, including the much-cited 'glider', led him to claim that in principle it could support the emergence of all recognizable animal forms and an infinite number of new ones. He also claimed that on a large enough scale there would be 'genuinely' living configurations, 'whatever reasonable definition' of living was applied (in Levy, 1992: 58). Among researchers and enthusiasts, CAs are deemed 'sufficiently complex to develop an entire universe as sophisticated as the one in which we live' (1992: 58). This sense of boundless 'creative' (Dawkins, 1991) if not creationist possibility persists in the development of alife software such as Richard Dawkins's *Biomorph* (Dawkins, 1991) and Thomas Ray's *Tierra* (Ray, 1996), and by means of the development of genetic algorithms (GAs) which are computer programs designed to replicate, mutate and evolve (Risan, 1996; Mitchell and Forrest, 1997).

11. Emergence is one of the key criteria for life outlined in the discourse of artificial life. It is linked with the concept of bottom-up or self-organization (Boden, 1996) as well as evolution and may be said to describe the capacity of the system (organic/informational) to develop behaviours or characteristics beyond those of its component parts. N. Katherine Hayles distinguishes between first-order emergence which refers to 'any behaviour or property that cannot be found in a system's individual components' and second-order emergence which signifies the appearance of a behaviour which stimulates the development of adaptive behaviours (1999: 9). Second-order emergence, then, involves the evolution of the ability to evolve and is one of the goals of alife. It is also what characterizes *Galápagos* as a metamorphic system.

12. In his seminal paper on artificial life (Boden, 1996) Langton gestured at ethical questions which remained largely dormant until a retrospective in 2001 (Bedau et al., 2001). Here, four main ethical issues are indicated: (1) the sanctity of the biosphere; (2) the sanctity of human life; (3) responsibility towards new forms of artificial life; and (4) the risk entailed in exploring the possibilities of alife. It is interesting that all these issues are similarly indicated in the discourse of genetics (the link is explicit in Bedau et al., 2001), suggesting that alife and genetics are informed by a single, fundamentally humanist bioethics which functions, if it functions, somewhat retrospectively as a brake on evolutionary possibility. In *Galápagos* there is no brake, no dialogue between the possible and the actual which, for Jacob is indicative of science (versus modern Darwinism), politics, ethics and the imagination. Without this, anything (and nothing) is possible, including the 'artificial selection' (Dawkins, 1991) of organisms based on purely aesthetic criteria.

13. In *Information Arts*, Stephen Wilson outlines the interactive and creative possibilities of alife and genetic art. Alife techniques 'can enable artists to create

interactive artworks beyond the simple-minded menu choice interactivity that characterises much multimedia' and the 'focus on developing believable synthetic worlds and creatures' can obscure the distinction between scientists and artist completely (2002: 341). Wilson anticipates a level of interactivity which enhances human/machine communication ('This may finally be a cybernetic ballet of experience, with the computer/machine and the viewer/participant involved in a grand dance of one sensing and responding to the other' [2002: 344]) and is predicated, as a 'cybernetic ballet of experience' on hybrid epistemologies and ontologies – cyborg/posthuman – which raise ethical and political questions alongside aesthetic ones.

14. See Dorothy Nelkin (1999).

15. *Genesis* was commissioned by *Ars Electronica 99* and presented online and at the O.K. Centre for Contemporary Art, Linz, from 4 to 19 September 1999.

16. See Mary Midgley (1992) and (2002).

17. Kac points out that: 'Morse code was chosen because, as first employed in radiotelegraphy, it represents the dawn of the information age – the genesis of global communications' (2003: 1).

18. See Kac, *Transcription Jewels* (2001).

19. Information, 'epitomised in the digital code or program' is for Langton similarly ' " implementation independent" – autonomous and divisible from material substrate' (Jonson, 1999: 48).

20. 'Transgenic art, I propose, is a new art form based on the use of genetic engineering techniques to transfer synthetic genes to an organism or to transfer natural genetic material from one species to another, to create unique living beings' (Kac, 1999: 289).

References

Bedau, M., J.S. McCaskill, N.H. Packard, S. Rasmussen, C. Adami, D.G. Green, T. Ikegami, K. Kaneko and T.S. Ray (2001) 'Open Problems in Artificial Life', *Artificial Life* 6: 363–76.

Boden, M.A. (ed.) (1996) *The Philosophy of Artificial Life*. Oxford: Oxford University Press.

Braidotti, R. (2002) *Metamorphoses: Towards a Materialist Theory of Becoming*. Cambridge: Polity Press.

Brown, A. (1999) *The Darwin Wars: The Scientific Battle for the Soul of Man*. London: Simon and Schuster UK.

Dawkins, R. (1976) *The Selfish Gene*. Oxford: Oxford University Press.

Dawkins, R. (1991) *The Blind Watchmaker*. London: Penguin.

Deleuze, G. (1994) *Difference and Repetition*, trans. P. Patton. London: Athlone.

Doyle, R. (1997) *On Beyond Living: Rhetorical Transformations of the Life Sciences*. Stanford, CA: Stanford University Press.

Fifield, G. (2000) *Art by Natural Selection*, http://www.genarts.com/galapagos/fifield97.html

Franklin, S., C. Lury and J. Stacey (2000) *Global Nature, Global Culture*. London: Sage.

Frauenfelder, M. (2003) *Do-It-Yourself Darwin*, wysiwyg://40/http://www.wired.com/wired/archive/6.10/Sims_pr.html

Gould, S.J. (2000) 'More Things in Heaven and Earth', in H. Rose and S. Rose (eds) *Alas, Poor Darwin: Arguments against Evolutionary Psychology.* London: Jonathan Cape.

Haraway, Donna J. (1989) 'Metaphors into Hardware: Harry Harlow and the Technology of Love', in D.J. Haraway *Primate Visions: Gender, Race and Nature in the World of Modern Science.* London: Routledge.

Haraway, Donna J. (1997) *Modest_Witness@Second_Millennium. FemaleMan©_ Meets_OncoMouse™.* London: Routledge.

Hayles, N. Katherine (1999) *How We Became Posthuman: Virtual Bodies in Cybernetics, Literature, and Informatics.* Chicago, IL: University of Chicago Press.

Hayles, N. Katherine (2001) 'The Condition of Virtuality', in P. Lunenfeld (ed.) *The Digital Dialectic: New Essays on New Media.* Cambridge, MA: MIT Press.

Jacob, F. (1982) *The Possible and the Actual.* Seattle: University of Washington Press.

Jones, S. (1999) 'Introduction', in James D. Watson *The Double Helix.* London: Penguin.

Jonson, A. (1999) 'Still Platonic After All These Years: Artificial Life and Form/Matter Dualism', *Australian Feminist Studies* 14(29): 47–61.

Kac, E. (1999a) 'Transgenic Art', in G. Stocker and C. Schopf (eds) *Ars Electronica 99: Life Science.* New York: Springer Wien.

Kac, E. (1999b) *Genesis,* http://www.ekac.org/geninfo.html

Kac, E. (2001) *Transcription Jewels,* http://www.ekac.org/genprot.html

Kac, E. (2002) *Transgenic Works,* http://www.ekac.org/transgenicindex.html

Kay, L. (2000) *Who Wrote the Book of Life? A History of the Genetic Code.* Stanford, CA: Stanford University Press.

Keller, E.F. (1995) *Refiguring Life: Metaphors of Twentieth-Century Biology.* New York: Columbia University Press.

Kember, S. (2002) *Cyberfeminism and Artificial Life.* London: Routledge.

Langton, C. ([1989] 1996) 'Artificial Life', in M.A. Boden (ed.) *The Philosophy of Artificial Life.* Oxford: Oxford University Press.

Lévy, P. (1998) *Becoming Virtual: Reality in the Digital Age,* trans. Robert Bononno. New York: Plenum Trade.

Levy, S. (1992) *Artificial Life: The Quest for a New Creation.* London: Jonathan Cape.

Maes, P. (1996) 'Artificial Life Meets Entertainment: Lifelike Autonomous Agents', in Lynn Hershman Leeson (ed.) *Clicking In: Hot Links to a Digital Culture.* Seattle, WA: Bay Press.

Maes, P. (1997) 'Modelling Adaptive Autonomous Agents', in C. Langton (ed.) *Artificial Life: An Overview.* Cambridge, MA: MIT Press.

Meek, J. (2001) 'ANDi, First GM Primate: Will Humans Be Next?', *The Guardian* 12 January.

Midgley, M. (1992) *Science as Salvation.* London: Routledge.

Midgley, M. (2002) *Evolution as a Religion.* London: Routledge.

Mitchell, M. and S. Forrest (1997) 'Genetic Algorithms and Artificial Life', in C. Langton (ed.) *Artificial Life: An Overview.* Cambridge, MA: MIT Press.

Monod, J. (1974) *Chance and Necessity.* London: Collins.

Nelkin, D. (1999) 'The Gene as a Cultural Icon', in G. Stocker and C. Schopf (eds) *Ars Electronica 99: Life Science*. New York: Springer Wien.

Oyama, S. (1985) *The Ontogeny of Information: Developmental Systems and Evolution*. Cambridge: Cambridge University Press.

Ray, T.S. (1996) 'An Approach to the Synthesis of Life', in M.A. Boden (ed.) *The Philosophy of Artificial Life*. Oxford: Oxford University Press.

Risan, L. (1996) 'Artificial Life: A Technoscience Leaving Modernity?', unpublished thesis.

Rose, H. and S. Rose (2000) 'Introduction', in H. Rose and S. Rose (eds) *Alas, Poor Darwin: Arguments against Evolutionary Psychology*. London: Jonathan Cape.

Rose, S. (1997) *Lifelines: Biology, Freedom, Determinism*. London: Allen Lane.

Shields, R. (2003) *The Virtual*. London: Routledge.

Sims, K. (1997) *Galápagos*, http://www.genarts.com/galapagos

Sims, K. (2003) *Decordova Exhibit Text*, http://www.genarts.com/galapagos/decordova-text.html

Todd, S. and W. Latham (1992) *Evolutionary Art and Computers*. London: Academic Press.

Unger, M. (2003) *Taking Over the Joystick of Natural Selection*, http://www.genarts.com/galapagos/nyt-unger99.html

Virilio, P. (2000) *The Information Bomb*. New York: Verso.

Warner, M. (2002) *Fantastic Metamorphoses, Other Worlds*. Oxford: Oxford University Press.

Warner, M. and S. Bakewell (2002) *Metamorphing: Transformation in Science, Art and Mythology*. London: Science Museum.

Wilson, S. (2002) *Information Arts: Intersections of Art, Science, and Technology*. Cambridge, MA: MIT Press.

Sarah Kember teaches in the Department of Media and Communications at Goldsmiths College, University of London. She is the author of *Cyber Feminism and Artificial Life* (Routledge, 2003) and other works in the field of feminist science and technology studies.

Making Music Matter

Mariam Fraser

A concert is being performed tonight. It is the event. (Deleuze, 1993: 80)

I

BRUCE GILCHRIST'S artEMERGENT project exploits 'raw neuro-physiological material' (www.artemergent.org.uk) for a variety of art processes and performances, among them (and perhaps best exemplified by) the *Thought Conductor* pieces. In the second of these, *Thought Conductor # 2* (*TC2* from hereon), the signals generated by an individual hooked up to an electroencephalogram are converted, via a relational-database devised by Johnny Bradley, into 'musical score'. This score appears on computer monitors, ready to be played by a string quartet. During the performance, things-to-see and things-to-hear manifest themselves discontinuously in space, as well as in a counter-intuitive temporal order: the graphic line of the EEG for instance, projected on a wall behind the stage, continues to move when the sound has stopped, while the score disrupts the chronological order of composition by unfolding *during* the performance – it is not completed until after the playing is over. Unable to locate the exact 'origins' of what there is to see and hear, or to anticipate where or how seeing and hearing will end, the observer experiences an enjoyable sense of perceptual disorientation.[1]

With passages of musical notation called up in the real time of the performance, score and sound acquire an immediacy which is characteristic of neuroscientific imaging technology in general, but which in this specific context lends new meaning to the notion of a 'live' performance. But let me be clear from the outset: *TC2* is not, in my view, a techno-scientific portrait that seeks to capture the 'inner kinetic melody' of the individual who sits at the centre of the stage. Of course, it is tempting to view it in these terms. The 'compère' who introduces the pieces, for example, describes them, rather reductively, in terms of the causal relations between the entities on the stage (and note that he begins with the brain):

> Here's how it works: the brain waves are picked up by this encephalograph machine, which is transmitted along this wire along here, along to a computer over here at the side of the stage which will translate it into music information which goes back along this wire onto the screen here, these notes are then picked up by the eye of the pianist, it goes along his optical nerve, down his spinal cord, along his arm and eventually press [*sic*] the notes here. (*Thought Conductor # 1*)

This emphasis on how the entities on the stage are physically connected to each other (by cables and wires and nerves) displaces the more significant questions, I think, which concern the connections between the 'author'/composer (if there is one), the scores,[2] and the sound. Importantly, the sound of *TC2*, like all 'non-measured rhythm', Jean-François Lyotard suggests, 'demands that one wait: what is happening?' (1991: 169). What is happening, or rather what *kind* of happening this is, will be the focus of this article. Throughout, I will be exploring the relations between the author, the score, and the sound in *TC2* as well in John Cage's *4'33"* (1952), the landmark work that inspired the *Thought Conductor* pieces, and which has been described as one of the first happenings in America. These key elements offer clues, as I will illustrate, to the different kinds of 'presences' that these performances embody, and the different notions of 'life' to which they refer.

II

4'33" was inspired by John Cage's visit to an anechoic chamber at Harvard University in 1951. Although he anticipated total silence, Cage instead heard two sounds: his own nervous system and his blood circulating. The visit marked a turning point in Cage's conception of sound and music, because it illustrated to him that 'pure silence is physically impossible' (Kostelanetz, 1991: 108). On this basis, he redefined silence as the absence of intended sounds, and music as 'an attentiveness to the sheer immediacy of an absolutely contingent conjunction of incidental sounds' (Pepper, 1997: 33). *4'33"* embodies these novel understandings. Composed according to chance principles, the three movements of silence – 'or in any event musically empty time' (Pepper, 1997: 33) – leave plenty of space for indeterminacy during the performance. Heinz-Klaus Metzger expands on the implications:

> It is a slap in the face of every traditional European aesthetic concept that the performance of Cage's work is a procedure largely constituted by accidents that are, strictly speaking, accidents of performance that cannot be related conclusively to notation. It is a further slap that during the performance the notations themselves refuse to generate a correlative sensuous appearance that would communicate meaning, since these notations are the results of mere chance operations in the technique of writing and in no way the formulations of a speaking subject. (1997: 5)

In this way, *4'33"* marks a break with what N. Katherine Hayles, drawing on Mark Poster, calls 'analogue subjectivity', subjectivity based on relations of resemblance. Analogue subjectivity, Hayles writes, is closely associated with print culture, and with the connection forged by alphabetic writing between a sound and a trace: 'to the extent that print can be considered an analogue medium, it connects voice to mark and thus author as speaker to the page' (1999: 14). This is not a spontaneous connection, as Michel Foucault shows in his analysis of the author-function. Instead, it is a consequence of the many operations (legal, psychological, aesthetic, etc.) 'that we force the texts to undergo, the connections that we make, the traits that we establish as pertinent, the continuities that we recognize, or the exclusions that we practice' (Foucault, 1991: 110).[3] These operations are no less relevant to the printed alphabet of music (notation) and to what I will call the author-composer. As Susan McClary notes, the word composition 'summons up the figure of a semidivine being, struck by holy inspiration, and delivering forth ineffable delphic utterances' (1999: 156). This is what *4'33"* challenges. In privileging chance and indeterminacy, composition is cast in terms of a kind of 'putting together' of materials that are decoupled from the laws of causality – from the laws that govern the relation between author and text, as well as the conventional structure of melodic plot 'in which the sound-matter is subordinated to a sentimental narration, an odyssey' (Lyotard, 1991: 173).

The success, or not, of Cage's desubjectivization strategies and his repudiation of organization have been discussed by critics and commentators at length. Some have suggested that Cage was far from abandoning the compositional subject, especially in his later pieces. Freed from 'the law of coerced labor as specified in the musical text and the conductor's baton', many of these scores confer on the musicians 'the dignity of autonomous musical subjects' (Metzger, 1997: 54). But even in *4'33"*, Liz Kotz argues, listening is an active invitation, 'inadvertently demonstrating the conservatism of this perceptual model, grounded in the express intentions of a centered subjectivity' (2001: 86). 'Cage', George Brecht said in an interview, 'was the great liberator for me . . . But at the same time, he remained a musician, a composer . . . I wanted to make music that wouldn't only be for the ears' (Brecht, in Kotz, 2001: 72). Although I will not be pursuing these particular critiques in any depth here, I do want to suggest – not entirely dissimilarly – that the displacement of the author-composer in *4'33"* is replaced by another kind of presence, that of 'life itself'.

As the title (a period of time on the clock) indicates, *4'33"* provides a temporal frame for non-intentional sounds. David Tudor, its first performer (at Maverick Concert Hall in Woodstock, New York in 1952), underlined the significance of the temporal structure of the piece when he commented that: 'the time was there, notated . . . except that the tempo never changed, and there were no occurrences – just blank measures, no rests – and the time was easy to compute. The tempo was 60' (Tudor, in Solomon, 1998). The time of the piece, in other words, parallels the ordered, systematic, and ostensibly unchanging temporal backdrop against which, since the late 19th century (Stengers with Gille, 1997), all other rhythms have been set and, by their deviance from this 'universal' tempo, quantified: clock time. Cage, Kotz writes, not only unhinged the score from '*sound* as a system of discrete notes but also from *time* as a graphically plotted system of rhythmic measure' (2001: 70). 'It is very important to read the notation,' Tudor said, '[i]t presents the impression that time is passing' (Tudor, in Solomon, 1998). Notably, this temporal frame has been likened to a space, a space-to-be-filled (in somewhat analogical fashion, one might say that time resembles space). It has been described, for example, as an 'airport for sounds', the musical companion to Rauschenberg's 'airports for shadows', the all-white paintings which greatly influenced Cage (Solomon, 1998). Homogenous space and time are set off from one another here then, but are nevertheless assumed to exist coincidentally. Let me spell out the implications.

As I have noted, Cage's redefinition of music, and *4'33"* in particular, served to '[erect] an absolute barrier' between writing, on the one hand, and sound, on the other (Pepper, 1997: 34). In an illuminating analysis of the implications of this distinction, Kotz explores how the model of the score as an independent graphic/textual object facilitates its 'rapid circulation between performance, publication, and exhibit formats: small, strange, and belonging to no definable genre, [it] could go anywhere' (2001: 60). However, if the score (the writing dimension of music) becomes newly mobile, the performance itself (the sound dimension) is all the more firmly rooted to a particular spatio-temporal context. The sounds of a particular performance of *4'33"* are usually (although not always) described in terms of its environment (perhaps the emphasis on the wind and the rain – 'environmental instruments' – during the first performance set the tone here). *At the time*, this is where its identity is located. The performance is a happening, because what happens, happens right here and right now. Cage's are the 'readymade sounds' to Duchamp's readymade objects (Maciunas, in Kotz, 2001: 80). 'What really pleases me in that silent piece', Cage said, 'is that it can be played any time, but only comes alive when you play it' (Cage, in Solomon, 1998).

The piece comes 'alive' when it is played because, as Cage understands it, the listener is alerted to the sound of the 'alive-ness' of a specific place at a specific time. In this respect *4'33"* seems to exemplify 'Cage's fascination', as one of his students described it, 'with the various theories of impersonality, anonymity and *the life of processes outside of their perceivers, makers or anyone else*' (Higgins, in Kotz, 2001: 64, original emphasis omitted, my emphasis added). It also indicates, however, that the distinction between writing and sound that Cage erects in *4'33"* is subtended by 'a poetics of presence, a mysticism of immediacy' (Pepper, 1997: 38). Now, it is the presence of presence itself, a particular kind of presence – in which time, space, and happenings all correspond with each other in a kind of self-identical one-ness/immediacy – that replaces the author-composer,[4] and from which the piece derives its *authority*.[5]

I want to suggest that the sound of *TC2*, by contrast, is the sound of an event. In the following two sections I will be using the terms pattern and event, which I have borrowed (rather loosely) from Alfred North Whitehead's theory of the organism,[6] to argue that the identity of *TC2* depends not on spatial relations of correspondence, resemblance, and/or distinction in time but, rather, on the organization of patterns *out of which* the regularities of space, time, and indeed analogue, emerge. *TC2*, in short, refers to a different kind of presence – although it is no less a 'happening' for that.

III

A musical score, even if it is not understood in terms of a blueprint for a performance, is usually assumed to be a relatively stable point of reference within it. This is certainly the case with *4'33"* which, although only minimally notated (to allow for a maximum of contingency of sound), does nevertheless *have* a score and can be rehearsed. It includes, for example, annotated recommendations for the performance, one of which advises the musician to use a stopwatch to time the movements. The completed score of *TC2*, by contrast, can only be read *after* the performance is over – necessarily so, because it is (or at least at first glance appears to be) the consequence *of* it.

Of course, no musical performance can be reduced to the score that it is (usually, conventionally) supposed to represent. On one level, this is a rather banal observation, since 'every piece of music we hear contains sound both intentional and non-intentional . . . no musical piece can twice give us exactly the same aural experience' (Kostelanetz, 1991: 108). *4'33"* and *TC2* are both distinguished in this regard, however, insofar as they are designed to draw attention to contingency and indeterminacy. In the case of *4'33"*, as noted above, the author-composer of contingency is the (life of the) space and time of the performance. In *TC2* it seems to depend (initially anyway) upon the inner bodily movements of the individual who takes centre stage/who is hooked up to the electroencephalogram. Indeed, part of the impact of *TC2* derives from the dramatic immediacy that is and has been associated (not necessarily correctly, as I will show below) with imaging technologies from Etienne-Jules Marey's early efforts to capture 'movement's signature, its rhythms and variations, in the form of graphic lines' (Lury, 1999: 505) to the PET scans of the present day. The graphic lines of *TC2*, if it was understood in these terms, would be the EEG reading (which is projected on a screen at the back of the stage), and the musical notation (the score, to which the EEG readings ostensibly give rise). By playing these lines, *TC2* could be said to be adding 'direct sound' to Marey's attempt to force 'direct writing' (Lury, 1999: 505, references omitted) from movement.

This is an 'intuitive' reading of *TC2*.[7] It is also, I would argue, a gravely mistaken one. It is mistaken because it assumes that, as in *4'33"*, the performance is designed to capture something that exists independently of it: the 'life of processes' in *4'33"*, and the 'life' of the individual in *TC2* (or more accurately and more modestly, neural activity). While there is reason to believe that this is the case in *4'33"* – on account of the score, which provides a frame that can be laid over a slice of life precisely *because* life is assumed not only to exist within the performance, but to proceed regardless *of* it – the *absence* of a score in *TC2*, at least until the end of the performance, signals that something different is at stake. Unlike the score of *4'33"*, which directs the musician as to how to capture sounds that exist *in* time during the performance, the score of *TC2* is the product *of* the time of the performance during which sounds are created. This is not simply a reversal of the temporal order, with notation now corresponding to sound. Instead, this temporal leap-frogging is an indication that, even though there *is* a score, the relation between notation and sound cannot and should not be cast straightforwardly in terms of analogical correspondences or resemblances at all. The reason is simple: the writing and sound dimensions of the 'music', as I will illustrate momentarily, are inextricably enfolded in each other. As such, they can neither be said to correspond to, nor resemble each other (for these notions assume a relation between two *separate* entities), nor could they be distinguished (a barrier could not be erected between them). One might say that they are as inextricable from each other as are elements in a pattern, where to remove an element would be to change the pattern. I want to suggest that it is not the performance as temporal frame – it is not *time* – that 'contains' the pattern; rather, it is the performance of *TC2* as event. I am redefining 'presence', in other words, in terms of 'some particular pattern as grasped in the unity of a real event' (Whitehead, 1985a: 130).

The term 'pattern' is helpful here, for it is a reminder that there are no things *qua* things that are grasped in an event, only *aspects* of things:

> The things which are grasped into a realised unity, here and now, are not the castle, the cloud, and the planet simply in themselves; but they are the castle, the cloud and the planet from the standpoint, in space and time, of the prehensive unification.[8] In other words, it is the perspective of the castle over there from the standpoint of the unification here. It is, therefore, aspects of the castle, the cloud, and the planet which are grasped into a unity here. (Whitehead, 1985a: 87)

A pattern cannot be captured or 'framed' therefore, either in its entirety or in isolation, since the very perception[9] of a pattern implicates the perceiver within it.[10] In this 'doctrine of mutual immanence . . . each happening is a factor in the nature of every other happening' (Whitehead, 1977: 41). Consider this, at a gross level, in relation to the performance of *TC2*. Here, the 'happening brain' (for example) which is ostensibly the original 'source' of the sound on stage also hears sound, and in hearing it is partially constituted by it, and by the other patterns in which the sound is implicated. Among those other patterns are the musicians who play what the machines read from the perpetual flux of sensory information in the brain and the machines which 'hear' the intelligence, fluency and emotionality of the playing which modifies that flux (and these are only the crudest of examples). This is no 'determinist sequence' in other words, but a 'relay of activities reacting the one upon the other, such that the actual occasion [the grasped pattern] itself in turn "conditions" the convergent series, and the convergent series the divergent ones, etc.' (Toscano, 2002). Rather than understand the score to be a *consequence* of the performance therefore (as I did above), or as the *product* of the (causal) relations between the staged entities (as an intuitive but, I have suggested, ultimately misguided interpretation implies), it can instead itself be considered to be a pattern that is set within, presupposes, and affirms other patterns.

Three fundamental implications follow from this conception of *TC2*. First, given the radical relationality that constitutes an event, it would be impossible to conceive of any aspect of the performance existing independently *of* it. It would be impossible, in other words, for a temporal frame to simply harness an autonomous and pre-existing readymade sound. To be in or of the event is to be defined *by* it. From this follows a second implication, which is that there can be no single author of contingency (no matter how impersonal), and no single subject-listener. The audience, for example, is not listening to sounds that are 'composed' by 'life' as it exists *in* or passes *through* a particular space and time. Instead, they are themselves *part of* the patterns that constitute the sound. In this respect one might say that the entire event 'listens' to itself. Conversely, even though the audience sit on the other side – both spatially and temporally – of the score on the screen, they too, along with every other pattern in the event, must be credited as its 'composers'. The identity of the event, in short, is defined not by any one of its (individual) components (such as the author-composer), or even by the sum of its components (all that a musical performance involves). It lies, rather, in the singular *becoming-together* of reciprocal prehensions. This becoming is, according to Whitehead, creativity itself: 'that ultimate principle by which the many, which are the universe disjunctively, become the one actual occasion, which is the universe conjunctively' (1985b: 21).

This conception of an indivisible multiplicity is significant (and this is the third point), because it dislodges from *TC2* the kind of presence which *4'33"* implicitly assumes (that is, presence understood in terms of corresponding self-identical unities: one time, one space, one happening). For instance, although the immediate source of the sound is clear in *TC2* (instruments are being played by musicians), the sound of the author-composer, the 'source' of the score, as I have suggested, is not. '[A]ll unlocated sounds are enigmatic,' Steven Connor writes, but 'unlocated voices are particularly so' (1997: 213) on account of the (analogical) relation between speech and the subject. However, while 'unattributed sound' might, in cinema, be 'marked by doubt and menace' (Connor, 1997: 213), in the context of *TC2*, it serves, I think, as an invitation to *look to* and *listen out for* events which are not directly present. By drawing the listener away from the immediacy of the performance as it is being performed, *TC2* seems to be signalling that its sound pertains not solely to the individual who sits at the centre *of* the stage, but to a pattern that cannot be contained *by* the stage. Indeed, this is the point of a pattern/event, as Whitehead says: 'This unity of a pre-hension defines itself as a *here* and a *now*, and the things so gathered into the grasped unity have essential reference to other places and other times' (1985a: 87). Or to put it differently: the notion of a pattern enables a certain 'listening away' which does not at the same time jeopardize the integrity of the event. The performance of *TC2*, in short, is no less 'present' for not being rooted entirely in the one space and time. I will illustrate this point now.

IV

A substantial part of the creation of *TC2* took place in Norway, where Gilchrist invited composers to score for a string quartet 'live' in his studio, while at the same time recording them both on video and on an encephalo-gram. This is how the 'mechanics' of the piece works. Time codes on the EEG and video were synchronized, enabling Gilchrist to correlate passages of the composers' musical notation with stretches of their EEG recordings (see www.artemergent.org.uk for details of these recordings).[11] These inter-actions – 'a wealth of musical output possibility' – were compiled within a database ('a creative engine') (www.artemergent.org.uk). During the live performances, Johnny Bradley's computer software program, called 'the DreamEngine', analyses the real-time EEG output from the individual on the stage, making matches (as closely as possible) with the contents of the database, and sending out musical notation to the members of a string quartet.[12] Insofar as the compositions that were scored in advance of the live performance bear an analogical resemblance to those produced during it, the process might be characterized in terms of what N. Katherine Hayles calls 'the Oreo' (after the black-and-white biscuit), an analogue–digital–analogue structure which she suggests often connects different embodied materialities.

In position emission tomography (PET), for example, a patient ingests radioactive substances whose decay is sensed by an instrument using analogue proportionality. These results are then inscribed as numerical data, digitally analysed, and manipulated 'to create lifelike analogical resemblances' (Hayles, 1999: 19). This resemblance ensures that the final images 'are often read as showing thinking in action' even though, as Hayles notes, 'they may more accurately be described as showing the Oreo effect in action' (1999: 19).[13] Similarly, while Gilchrist's correlation of the composers' passages of composition with their EEG readings looks like a celebration of analogue subjectivity in action – wherein, as Hayles describes it, 'what is at the forefront of the mind is also imagined to be deep inside' (1999: 13) – in fact the scores that are produced during the perform-ance are not the same as those written by these composers. The 'detour' through the digital ensures that the break between the interiority of the composer and the score – one of the most challenging aspects of *4'33"* – is maintained in *TC2*. It also arguably illustrates what Hayles identifies as one of the 'distinctive advantages' of the digital aspect of the Oreo structure. As she puts it:

> Moving from analogue resemblance to coding arrangements opens possi-
> bilities for leveraging unthinkable with analogue resemblance, which by
> virtue of *being* a resemblance must preserve proportional similarity . . .
> Coding arrangements have powerful transformative properties precisely
> because they have been freed from the morphological resemblances of
> analogue technologies. (1999: 19)

And, indeed, by passing the 'original' composers' scores through the
database, *TC2* exemplifies not only the transformative, but what Lyotard
calls the 'liberatory', potential of technology. Like Cage, Gilchrist and
Bradley abandon rhythmic measure for mechanical clock time: although the
composers scored in various tempos, once transcribed into the database
'everything becomes events along a timeline (milliseconds)' (Gilchrist,
personal correspondence). The 'metronomy of sound-time,' as Lyotard
describes it, 'the discrimination of duration in classical notation by breve,
semibreve, crochet, quaver, semiquaver, etc.' is replaced here by 'the
continuous race of the chronometer' (1991: 168–9). This shift to chrono-
metric time enables Gilchrist to start and stop the 'music' several times
during a live performance, as different individuals take the stage. Such
seemingly 'arbitrary' starting and stopping breaks with the model of cause
and effect implied by 'the dialectic of epic which encloses the time of the
work in a beginning, a development and an end – with its harmonic coun-
terpart, resolution' (Lyotard, 1991: 173). In other words, like Cage, Gilchrist
seems to want to 'free' 'the material – sound – from [at least some of] the
various constraints that it had to respect . . . in order to make itself musically "presentable"' (Lyotard, 1991: 168).

Although the digital phase offers many possibilities for fragmentation
and recombination, for Hayles, the return to analogue resemblance is
important. First, she claims, because it is 'the mode best suited to our
sophisticated visual-cognitive perceptual skills' – and here she compares
the instantly intuitive accessibility of the PET image to an array of numerical data which may take hours or even days to unravel – and, second,
because it is 'the dynamic that mediates between the noise of embodiment
and clarity of form' (1999: 19). Hayles makes this latter point in the context
of what she calls 'the material/discursive divide' that informs the work of,
in her words, 'scientific realists':

> For the realist, the flow of structuring information about physical reality
> moves from the material (say, a field of morning glories . . .) through the oper-
> ational (experiments in breeding that operate upon the plant . . .) to the
> symbolic (graphs and charts . . .). The closer the researcher is to the embodied
> reality of the plants, the fuzzier the picture is likely to be as various sources
> of 'noise' and 'contamination' complicate the regularities presumed to be
> revealed by such inscriptions and charts. (1999: 17)

4'33" is arguably 'experimental' in a 'realist' sense: it uses the score to
'operate' on 'the material', that is, to isolate material sounds in time.[14] This

is why, as I argued earlier, the performance is immobile: in this conception of the world, '[m]aterial embodiments do not circulate effortlessly' (Hayles, 1999: 17) – indeed a performance of *4'33"* cannot, by definition, circulate at all. The score, however, as I also noted, can and does circulate (see Kotz, 2001, on the implications of this mobility). Again, this is enabled by a particular conception of the world in which 'cultural conventions privilege the forms expressed by the inscriptions over their instantiations in particular media . . . which are regarded as passive vehicles for the transmission of the forms' (Hayles, 1999: 17. See Lury [2004], who challenges this notion in the context of brands).

My understanding of the score of *TC2* in terms of a *pattern* complicates these distinctions.[15] *As* a pattern – which contains within it, and is contained by, other patterns – the score too must be understood to be a 'concrete' entity which is specifically instantiated in specific contexts. '"Actuality" means nothing else than this ultimate entry into the concrete,' Whitehead writes, 'in abstraction from which there is mere nonentity' (1985b: 211). *TC2* is compelling because it seems to purposely draw our attention to this point. The audience understands the score to be immanent to the performance – immersed in all the occurrences that take place during it – because we literally see it appear *during* the performance. The score embodies the performance, and is embodied by it. In this respect it is clearly *not* a passive vehicle for the transmission of symbolic marks or forms, or a transparent conduit for the author-composer's intentions. It is no less 'noisey' than the material sounds themselves; indeed, as a pattern contained within and containing other patterns, it can be understood to embody those sounds within it.

This is not to suggest, however, that the score is solely the product *of* the performance. For although aspects of *TC2* might be *understood* in terms of analogical resemblance, in fact the audience does not see – and may not even imagine that there were – 'original' compositions, written in Norway, which the 'final' scores resemble (as we have become used to imagining a brain which a PET image is assumed to resemble). The audience does not, in other words, see the top part of the Oreo structure during the live performance. However, even if one *were* aware of the 'original' compositions, the *sonic* dimension of *TC2* cannot be said to bear any resemblance to anything at all.[16] Unlike the 'original' scores, which bear an analogical relation to the ones that are created during the performance, the *sound* of the composers' 'original' compositions remain unactualized prior to the performance. There simply is no sonic 'bottom' part of the Oreo. It is for this reason – because the challenge of *TC2* is lodged through sound as well as through a (visual) score – that the analogical relations between the author-composer, the score, and the sound (whereby the performance is the 'author' of the score, which corresponds to the sounds created during it) cannot be reinstalled. There is no author-composer, as I have argued, but nor do the sounds correspond to the performance/the performance-score.

Does this mean that the composition that is created during the performance is *absolutely* original, an ostensibly 'unauthored' creation, a creation *ex nihilo*? The model of creativity that I have drawn on for the most part in this article would certainly imply as much, for it seems that there must always be an author of some kind, no matter how 'impersonal'. However, I have also noted that there are other ways of conceiving of creativity, and that Whitehead's conception of creative activity, as 'the process of eliciting into actual being factors in the universe which antecedently to that process exist only in the mode of unrealized potentialities' (1977: 26–7), is best suited to *TC2*. In fact, this is a particularly fitting understanding insofar as the 'pure' sonic object, like potentiality (or even: the pure sonic object *as* potentiality), has no physical or spatio-temporal existence outside of its specific actualization. It has no presence, in the *4'33"* sense. This is why it cannot be contained within the Oreo structure that Hayles describes. Thus, I will define creative activity, in the context of *TC2*, as the process by which the potential sonic object is actualized, in different ways and at various different points, during the live performance. This is how the (unactualized) sounds of the scores that were written (and EEG readings that were taken) in Norway are best understood: in terms of a reservoir of potentiality, rather than as a template or a meta-composition. Nevertheless, while the elicitation of potentiality always amounts to something new insofar as it is a novel concrescence of disjunctive diversity (see my earlier discussion of Whitehead's notion of creative activity), for this very same reason it cannot be said to be 'new in the sense of being completely different elements from that of the past' (Halewood, 2003: 127).

So what are the implications of this revised understanding of creativity, in relation to *TC2*? In the first instance it is worth noting that although a particular performance of *TC2* can never be repeated (it is always a novel becoming), this is not because it is bound down by, or solely referenced to, the happenings that happen in *that* particular space and time. On the contrary: 'There is time because there are happenings, and apart from happenings, there is nothing' (Whitehead, 1920: 66). Space and time are not autonomous entities *in which* (patterns of) sounds are situated in other words, but are rather abstractions generated *by* overlapping, interactive, patterns.[17] Thus, it is that the potentiality that is actualized during a performance of *TC2* cannot be said to *precede* the performance in any linear temporal sense. The sounds of *TC2* are not the realization of some kind of nascent form, for example. Instead, space and time are the forms, or regularities as I would rather put it, that emerge out of patterns. They are not universal laws that can help us to understand them.

This same might be said of the analogue and the digital – that is, that they emerge out of particular events, rather than describe them (as Hayles herself agrees, in somewhat different terms [1999: 23–4]). It does not make sense therefore, to privilege digitization as *the* site of transformation.[18] Creative, differentiating, activity is everywhere on-going. It is on-going, and also limiting. Necessarily so, for it would clearly be impossible for the entirety of potentiality to be actualized during a performance.[19] Creative activity must thus be understood to refer not only to what is *included* in a pattern, but also to 'the exclusion of the boundless wealth of alternative potentiality' (Whitehead, 1977: 27). Finally, then, it is this emphasis on what is *not* 'there', as well as on what 'is', which indicates that presence and im/mobility are not really the best terms in which to conceive of an event. Instead, a pattern endures to the extent that it prehends and is prehended by other patterns, 'to the extent that the environment is a relevant aspect for it and vice versa' (Toscano, 2002). *TC2*, as an event, 'has relevance' (it endures) whenever and wherever its pattern is instantiated – during a performance for example, but also here and now, as I am writing this article and as you are reading it.[20]

V

To sum up. I mentioned Brecht's critique of Cage in passing earlier. I want to cite it again here, more fully:

> Cage . . . was the great liberator for me . . . But at the same time, he remained a musician, a composer . . . I wanted to make music that wouldn't only be for the ears. Music isn't just what you hear or what you listen to, but everything that happens. (Brecht, in Kotz, 2001: 72)

The reason that *4'33"* only 'comes alive' when it is performed is because it requires a listener to attend to the sound of 'life'. 'Art', Cage said in interview, 'is not an escape from life. It is an introduction to it' (Cage, in Solomon, 1998). This must be a life that sings in solo however, for its sound continues whether the subject is listening or not (hence Cage's fascination with theories of 'impersonality' and 'anonymity'). In *TC2*, by contrast, the distinction between the author-composer (life, in *4'33"*) and the listener, between the individual and the 'external' environment – between any discrete entities in fact – cannot be clearly maintained. Music and ears and listening and . . . (everything) *require* each other as patterns within patterns over-lapping patterns require each other. In this sense music is indeed, as Brecht says, 'everything that happens'.[21]

There is indeterminacy and contingency in *TC2*, but it is not the indeterminacy and contingency of sounds that are accidentally/unintentionally captured *in* time, but of the actualization *of* sound *and* time out of potentiality. And there is originality, but not the originality of original scores being played for the first time, but of novel patterns becoming-together for the only time. This, for Whitehead, is life: 'a single occasion is alive when . . . its process of concrescence has introduced a novelty of definiteness not to be found in the inherited data . . . [Life] is the name for originality, and not for tradition' (1985b: 104). I will conclude this article with a rather dramatic flourish therefore, and suggest that *TC2* does not introduce the listener *to* life, but that it is *itself* alive, and that the life of the listener is a part/pattern of that.

Coda

The pattern of *TC2* is both in and out of 'science', in and out of 'critique', and in and out, I would argue, of 'art'. Indeed, its pattern undoubtedly contributes to the emergence of these spheres, for which it shows no respect. But this is precisely the point, as Deleuze (Deleuze and Parnet, 1987) long maintained – to be *and* rather than either/or. And it is perhaps particularly important to be *and* in a context where, as Stengers wryly puts it, the very distinction between '*either* objective, neutral, having the power to disconnect itself *or* mere construction . . . is a way for scientists to obtain limitless patience [relevance, in the terms of this article] from their environment, where some "impatience" should well prevail' (2002: 250). I too will be impatient and claim that the evaluation of *TC2* depends not on its *location* (a problematic term, which this article has grappled with at length), but on what its pattern continues to make relevant (or not), and what new relevances it establishes. This is how its value is to be defined, for like all of Whitehead's entities, it has no higher value than itself.

Acknowledgement

I am very grateful to Bruce Gilchrist for his generosity in discussing his work with me informally. Thanks also to Andrew Barry, Georgie Born, Mike Gane, and Celia Lury for their constructive comments on earlier versions of this article, and to Mike Featherstone for his continued encouragement. This piece is written for Steven W., for whom music matters.

Notes

1. Of course the experience of *TC2* necessarily involves far more than seeing and hearing. However, I will be confining my comments in this article to the relation between the visual and the aural – which I have further confined to the relations between the score (that can be seen) and the sounds (that can be heard).

2. Several scores are produced during the course of the performance, as several individuals are hooked up to the EEG machine. From now on, though, for the sake of clarity, I will be referring to just the one score/individual. It is also worth noting at the outset that although Gilchrist and Bradley could have printed out copies of the score during or after the performance, they did not choose to do so. Scores from the performance are archived as MIDI files.

3. See Lury, this volume, pp. 93–110 for an analysis of how the relationship between author and text has changed, and continues to change, particularly as the practices of art and science (which have, historically, differently mediated the author-function) begin to converge.

4. So what role does Cage play here, then? He could be described as a conductor, but I would prefer to identify him as an 'operator' (see below, where I compare *4'33"* to an experiment, and note 14).

5. And confers it: one of the anonymous reviewers of this article for instance emphasized that, to quote, 'I was present when Cage played his *4'33"* silence piece at Princeton in 1953.'

6. In this theory, Whitehead outlines his conception of 'concrete' things or, in his terms, actual entities or occasions. He also, sometimes, describes a nexus of actual

occasions in terms of an event. Understood thus, events can be identified at every level and in every register of life (cf. Stengers, 2000, who mostly confines her understanding of events to the macro-level).

7. One that would arguably situate it alongside other pieces which seek to externalize the 'interior' of the body for public scrutiny. Gunther von Hagens's 'spreads' immediately come to mind – although where von Hagens looks back to pathological anatomy (what Foucault calls 'the technique of the corpse'), Gilchrist might be more closely aligned with the vivifying gaze of physiology.

8. 'Every prehension,' Whitehead writes, 'consists of three factors: (a) the "subject" which is prehending, namely, the actual entity in which that prehension is a concrete element; (b) the "datum" which is prehended; (c) the "subjective form" which is *how* that subject prehends that datum' (1985b: 23).

9. Stengers writes: 'Mark well, not what we perceive and can identify, but the whole indefinite complexity of what we are aware of, even if we have no words to name it' (1999: 197).

10. Indeed, perception 'itself' will be a part of a pattern (as Cage's experience in the anechoic chamber illustrates). This is what Whitehead calls 'presentational immediacy', which is basically, as Judith Jones explains, 'an experienced display, by and for the percipient subject, of the present environment' (1998: 150). This does not mean that the percipient experiences 'bare sensations' which are 'then "projected" into [her] feet as their feelings, or onto the opposite wall as its color' (Whitehead, in Jones, 1998: 151). Instead, '[t]he projection is an integral part of the situation, quite as original as the sense-data' (Whitehead, in Jones, 1998: 151).

11. Correlating this material was necessarily dependent upon the working process of the composer. Often, Gilchrist notes,

> the composer would work on a bar, then skip back to a previous bar to add a little bit more . . . If a bar was revisited, I had the choice of lifting it out in an incomplete state, a kind of 'bar in progress'. (personal correspondence)

12. It would be possible to argue that the performance is limited by the size and complexity of the material in the database. At the time of writing, Gilchrist and Bradley conjecture that each 'individual piece' (of which there are usually about four or five during a performance) could last around 10 minutes before this material would begin to repeat itself. On the other hand, the notion that the performance is limited (and conversely, that with more material it could be really extensive) is something of a red herring, since nearly all databases are plagued by problems concerning size and complexity, to the extent that these problems could almost be said to be *definitive of* a database (see for example, on the archive, Alan Sekula, 1986, and George Myerson, 1998).

13. Note, as Hayles does, that although computers are often considered to be the digital dimension of the Oreo structure, these too have an analogue top and bottom that enables the human user to interact with its processes (1999: 19).

14. Perhaps the answer to the question of Cage's role, if 'life' is the author-composer, lies here: although not quite a scientist in the sense that Lury suggests Hirst might aspire to be (see Lury, this volume, pp. 93–110), Cage could certainly be described as an 'operator'.

15. See also Barry, this volume, pp. 51–69, for a related critique of the distinction

between information and materiality in the context of the pharmaceutical industry.

16. Of course it is possible to conceive of the analogue in precisely this way. Brian Massumi, for example, understands the virtual to be

> more analogical than descriptive. It is not, however, an analog *of* anything in particular. It is not an analog in the everyday sense of a variation on a model. Here, there is no model . . . The analog is *process*, self-referenced to its own variations. (2002: 135)

Although my own argument, particularly regarding potentiality (see below), overlaps with Massumi's in parts, it does not map directly on to it as his different understanding of analogue indicates.

17. In this respect I would argue that, where rhythmic measure is replaced by clock-time in *4'33"*, clock-time is replaced by event-time in *TC2* – or rather with duration, as Whitehead understands it. Duration, Whitehead writes, 'is the field for the realised pattern constituting the character of the event' (1985a: 157). A pattern is realized through the becoming temporal of a duration – '[t]emporalisation is realisation' (Whitehead, 1985a: 159) – and, in order to endure, requires a succession of durations (each one exhibiting the pattern), rather than a succession of durationless instants. It is not autonomous time which is divisible (into instants) therefore; instead, 'divisibility . . . is within the given duration' (Whitehead, 1985a: 158). In short, the divisibility and extensiveness of the pattern are derived from its own duration, and not from spatio-temporal relations that are perceived to be external to it. Note, for interest, that in *0'00"* (1962) (or *4'33" No. 2* as it is sometimes called) Cage drops measured time altogether.

18. Indeed, Massumi argues that digitization offers only limited opportunities for transformation insofar as it refers mainly to the sphere of the possible:

> Digitization is a numeric way of arraying alternative states so that they can be sequenced into a set of alternative routines . . . 'To array alternative states for sequencing into alternative routines.' What better definition of the combinatoric of the possible? . . . It doesn't bother approximating potential, as does probability. Digital coding is possibilistic to the limit. (2002: 137)

19. This would be like understanding memory in terms of something (like a score, for example) which is 'stored' and which can be called up and replayed absolutely. Consider, by contrast, Henri Bergson's conception of memory as a virtual presence of the past in the present, only a fraction of which is actualized at any particular moment. In a description of 'change' in memory which bears a striking resemblance to Whitehead's notion of creative advance, Bergson notes that although 'the continuous life of a memory . . . prolongs the past into the present' (1912: 44), it is also the case that 'the second moment always contains, over and above the first, the memory that the first has bequeathed to it' (1912: 12).

20. 'The one-ness of anything that exists', Judith Jones writes, 'is nothing else but its *multiple realization* in the universe of events, and thus its oneness, or self-sameness, is not established simply by the boundaries of its own becoming' (1998: 128).

21. As Whitehead puts it: 'We are in the world, and the world is in us' (1977: 42).

It is worth considering Steven Connor's description of 'the auditory self' here, even though he is writing in a very different context:

> The self defined in terms of hearing rather than sight is a self imaged not as a point, but as a membrane; not as a picture, but as a channel through which voices, noises and musics travel . . . The auditory self discovers itself in the midst of the world and the manner of its inherence in it . . . [It] is an attentive rather than an investigative self, which takes part in the world rather than taking aim at it. (1997: 207, 219)

References

ArtEMERGENT project, www.artemergent.org.uk, accessed June 2002.

Barry, A.W. (2005) 'Pharmaceutical Matters: The Invention of Informed Materials', *Theory, Culture & Society* 22(1): 51–69.

Bergson, H. (1912) *An Introduction to Metaphysics*, trans. T.E. Hulme. New York: G.P. Putnam's Sons.

Connor, S. (1997) 'The Modern Auditory I', in R. Porter (ed.) *Rewriting the Self: Histories from the Renaissance to the Present*. London: Routledge.

Deleuze, G. (1993) *The Fold: Leibniz and the Baroque*, trans. T. Conley. London: Athlone Press.

Deleuze, G. and C. Parnet (1987) *Dialogues*, trans. H. Tomlinson and B. Habberjam. London: Athlone.

Foucault, M. (1991) 'What is an Author?', in P. Rabinow (ed.) *The Foucault Reader: An Introduction to Foucault's Thought*. London: Penguin.

Halewood, M. (2003) 'Subjectivity and Matter in the Work of A.N. Whitehead and Gilles Deleuze: Developing a Non-Essentialist Ontology for Social Theory', unpublished PhD thesis, University of London.

Hayles, N.K. (1999) 'Simulating Narratives: What Virtual Creatures Can Teach Us', *Critical Inquiry* 26: 1–26.

Jones, J. (1998) *Intensity: An Essay in Whiteheadian Ontology*. Nashville, TN: Vanderbilt University Press.

Kostelanetz, R. (1991) 'Inferential Art', in R. Kostelanetz (ed.) *John Cage: An Anthology*. New York: Da Capo Press.

Kotz, L. (2001) 'Post-Cagean Aesthetics and the "Event" Score', *October* 95: 55–89.

Lury, C. (1999) 'Marking Time with Nike: The Illusion of the Durable', *Public Culture* 11(3): 499–526.

Lury, C. (2004) *Brands: The Logos of the Global Economy*. London: Routledge.

Lury, C. (2005) '"Contemplating a Self-portrait as a Pharmacist": A Trade Mark Style of Doing Art and Science', *Theory, Culture & Society* 22(1): 93–110.

Lyotard, J.-F. (1991) *The Inhuman: Reflections on Time*, trans. G. Bennington and R. Bowlby. Oxford: Polity Press.

Massumi, B. (2002) 'On the Superiority of the Analog', in *Parables for the Virtual: Movement, Affect, Sensation*. Durham, NC: Duke University Press.

McClary, S. (1999) 'Afterword: The Politics of Silence and Sound', in J. Attali *Noise: The Political Economy of Music*. Minneapolis: University of Minnesota Press.

Metzger, H.-K. (1997) 'John Cage, or Liberated Music', *October* 82: 49–61.

Myerson, G. (1998) 'The Electronic Archive', *History of Human Sciences* 11(4): 85–101.

Pepper, I. (1997) 'From the "Aesthetics of Indifference" to "Negative Aesthetics": John Cage and Germany 1958–1972', *October* 82: 31–47.

Sekula, A. (1986) 'The Body in the Archive', *October* 39: 3–65.

Solomon, L.J. (1998) 'The Sounds of Silence: John Cage and *4'33''*', http://www.azstarnet.com/~solo/4min33se.htm

Stengers, I. (1999) 'Whitehead and the Laws of Nature', *SaThZ* 3: 193–206.

Stengers, I. (2000) *The Invention of Modern Science*, trans. D.W. Smith. Minneapolis: University of Minnesota Press.

Stengers, I. (2002) 'Beyond Conversation: The Risks of Peace', in C. Keller and A. Daniell (eds) *Process and Difference: Between Cosmological and Poststructuralist Postmodernisms*. Albany, NY: SUNY Press.

Stengers, I., with D. Gille (1997) 'Time and Representation', in I. Stengers *Power and Invention: Situating Science*, trans. by P. Bains. Minneapolis: University of Minnesota Press.

Thought Conductor # 1 (video).

Thought Conductor # 2 (video).

Toscano, A. (2002) 'The Theatre of Production: Philosophy and Individuation between Kant and Deleuze', unpublished PhD thesis, University of Warwick.

Whitehead, A.N. (1920) *The Concept of Nature*. Cambridge: Cambridge University Press.

Whitehead, A.N. (1977) *Nature and Life*. New York: Greenwood Press.

Whitehead, A.N. (1985a) *Science and the Modern World*. London: Free Association Books.

Whitehead, A.N. (1985b) *Process and Reality*, revised edn, ed. D.R. Griffin and D.W. Sherburne. New York: Free Press.

Mariam Fraser is Lecturer in Sociology at Goldsmith's College, University of London. Her current research focuses on the relations between ontology, value, and ethics. Her forthcoming book, *The Value of Ethics*, explores these issues through an archival study of the development and production of the antidepressant Prozac. This project is funded by the Wellcome Trust Biomedical Ethics research programme (grant no. 065209/Z/01/Z/CM/CD/SW).

Index

Theory, Culture & Society

Theory, Culture & Society caters for the resurgence of interest in culture within contemporary social science and the humanities. Building on the heritage of classical social theory, the book series examines ways in which this tradition has been reshaped by a new generation of theorists. It also publishes theoretically informed analyses of everyday life, popular culture, and new intellectual movements.

EDITOR: Mike Featherstone, *Nottingham Trent University*

THE TCS CENTRE
The Theory, Culture & Society book series, the journals *Theory, Culture & Society* and *Body & Society*, and related conference, seminar and postgraduate programmes operate from the TCS Centre at Nottingham Trent University. For further details of the TCS Centre's activities please contact:

Centre Administrator
Theory, Culture & Society Centre
School of Arts, Communication & Culture
Nottingham Trent University
Clifton Lane, Nottingham, NG11 8NS, UK
e-mail: tcs@ntu.ac.uk
web: http://tcs.ntu.ac.uk